Quaker Sons Go to War

QUAKER SONS GO TO WAR

The Allinson Family of Burlington County, New Jersey

DOROTHY ROBBINS TALAVERA

South Jersey Culture & History Center

2026

Quaker Sons Go to War: The Allinson Family of Burlington County, New Jersey

By Dorothy Robbins Talavera

Copyright © 2026 Dorothy Robbins Talavera

Design and layout Copyright © 2026 the South Jersey Culture & History Center at Stockton University

Printed in the U.S.A.

stockton.edu/sjchc/

ISBN: 978-1-947889-25-5

This book is dedicated to the families and loved ones of all military service personnel who have ever been deployed: to those who carry on and wait, to those who mourn, and to those who live with who and what comes home.

Table of Contents

A disclaimer about what this book is not: it is not a scholarly treatment of the Civil War or military maneuvers; it is not an in-depth study of Quakerism; it does not contain definitive Allinson family biographies. Other authors, much more qualified than I, have already done an excellent job on those topics. If any of that is what you are seeking, you will need to return to the library. Quite frankly, until relatively recently, I knew very little about most of the people in this book.

So, what is this book? It is my journey to get to know long-gone ancestors who left an intriguing paper trail.

All my life, I have known about "the Civil War letters" in my mother's trunk of family archives. When I inherited that trunk, I found the letters and attempted to read them; it was slow going.

They were hard to read due to a lack of modern grammar, punctuation, and uniform spelling. Some had been turned 90° and written across the previous lines to save paper. So, I decided to transcribe them, organize them, and try to make sense out of what I had.

Initially, I just photocopied what I transcribed, had it spiral bound at Staples, delivered a booklet to each family member, and donated one to the Burlington County Library, and another to the Fort Ward Museum. That was in 2008. It has, however, been bothering me lately that the people in the letters deserve more. There is a saying that we die twice: once when our body passes, and again when our name is forgotten. That is what inspired me to say their names.

As I spent time with these letters, I began connecting names on the back of old photographs. I researched genealogical records, old newspapers, books, and other records and found myself becoming acquainted with these folks. I wanted to know more about their moral dilemma and the suffering they endured. I wanted to honor the sacrifice of these men and of those they left behind.

Readers will notice I provided information disproportionately about one brother. He is the only brother for whom I own primary sources in the form of letters he wrote. I describe two military units within these pages: the First New Jersey Infantry, and the Twenty-Third Regiment of New Jersey Volunteers. I have divided the book into two parts: Part 1 contains everything I know about the three Allinson brothers. Part 2 contains word-for-word transcription of Joseph's letters, along with explanations to place the contents into context.

Dorothy Robbins Talavera
Delanco, New Jersey
August 2025

PART I
THE ALLINSONS AND COOPERS:
An illustrious Quaker heritage

This is the story of three Burlington County, New Jersey, brothers who went to war, each in his own way. Joseph Allison/Allinson (1826–1862), David Cooper Allinson (1827–1908), and John Cooper Allinson (1828–1900) had a long pedigree, with ancestors going back to the earliest settlers in southern New Jersey, including some of the most prominent Quaker[1] families. In addition to the Coopers and the Allinsons, well-known New Jersey surnames such as Burr, Matlack, and Haines appear in the direct family tree.

The Cooper Family

William Cooper (1632–1710) is considered the Father of Camden, New Jersey. He had become a member of the Religious Society of Friends in his native England, where he met fellow Quaker William Penn. In 1679, Cooper sailed with his wife and five children to present-day Burlington, New Jersey, where they bought land and temporarily settled. A year or so later, Cooper bought 300 acres near the point where a river that now bears his name meets the Delaware River. He was present at the signing of William Penn's treaty with the Indians in 1682, continued to expand his properties, and became an important political leader.[2] Cooper's descendants continued to be leaders of the burgeoning community. William Cooper is buried in Burlington.

The Allinson Family

Another distinguished Quaker family emigrated to Burlington, arriving almost 40 years after the Coopers. Joseph Allinson (1692–1756), the original immigrant, left England and arrived in Burlington, New Jersey during 1718. There, he became a devoted member of the Burlington Monthly Meeting, established in 1678 where "weighty Quakers," wealthy men who were active in business and politics, surrounded him. Burlington had long served as a governmental and commerce center.

Joseph's son Samuel (1738–1792) became a licensed attorney. He became active in education, politics, and committed to social justice issues. He was among the first Trustees for the Friends' School in Burlington. In the mid-1700s, Samuel was falsely accused of charging exorbitant fees for real estate transactions. In his well-documented defense, Samuel turned the tables, ably proving his accuser (the Sheriff and a member of the Legislature) to be the person who committed that offense.[3] The New Jersey colonial government assigned Samuel the task of collating and publishing a comprehensive edition of the provincial laws.

Quakers grappled with the complex issues surrounding slavery during the late eighteenth century. Some wealthy Quakers owned slaves; members of one monthly meeting listed owning as many as 1,100 enslaved human beings,[4] and Samuel became active in efforts to end it. Even today, Samuel is known as the "Burlington Emancipator."[5] In 1774 he wrote a letter to Virginia patriot Patrick Henry, urging liberty for all. He wrote that

> . . . a fairer time never offered to give a vital blow to the
> shameful custom of Slavery in America . . . groaning
> under unconstitutional impositions destructive of their
> Liberty. Can we say that a limited Slavery is injurious
> & disagreeable to ourselves, &, by our practice declare,
> that absolute Slavery is not unjust to a race of fellow Men
> because they are black?[6]

Samuel and his first wife Elizabeth Smith (1737–1768) had three children. Following Elizabeth's death, Samuel married Martha Cooper (1747–1823) in 1773. Martha was a great-great-granddaughter of Camden founder William Cooper. Samuel and Martha had nine children. Samuel is buried with Martha in the Burlington Friends Meeting cemetery. He left behind a legacy of respect, admiration from his peers, and the expectation that his heirs would continue that path.[7]

One of Samuel's grandsons was William J. Allinson, (1810–1874), a skillful and influential orator, writer, and editor, who joined the abolitionist movement. Among other achievements, he promoted the growth of the Burlington City public library. He opened a pharmacy in Burlington. The Burlington Quaker Meeting House and Center for Conference lists William, editor of *Friends Review*, among the notable Quakers buried there.

David Allinson (1774–1858)[8] was the son of Samuel and Martha Cooper, uncle to William J. He continued the tradition of education and letters. In 1803, David began a printing and publishing career in the City of Burlington, New Jersey. David's works included the weekly *Rural Visiter* and the *Saturday Evening Visitor* newspapers, which were described as useful and entertaining. Newspapers all along the east coast advertised books he had for sale, books he had recently written and published, and notices that he was authorized to accept periodical subscriptions.

David was appointed a Trustee of the Burlington Academy, established in 1795.[9] His name appears several times when he signed public letters recommending teaching candidates: "We, the Trustees of the Burlington Academy . . . certify . . . that the bearer has for some years been a teacher at said Academy, and that we have always considered him fully competent . . . being well convinced that his knowledge of the Greek and Latin languages is such as few in this country have surpassed or equaled. . . ."[10]

In "meridian life," David experienced financial setbacks in his business, and became mired in debt.[11] "David Allinson has become so embarrassed in his trade of business as to be unable to discharge his debts on which account he has been visited by the overseers who are appointed to make inquiry into the case and report to our next meeting."[12] This led to the Religious Society of Friends disowning him in 1817. He then joined Saint

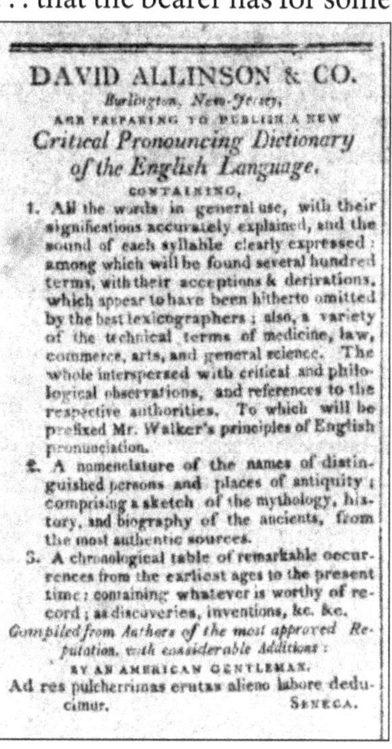

DAVID ALLINSON & CO.

Burlington, New-Jersey,

ARE PREPARING TO PUBLISH A NEW

Critical Pronouncing Dictionary of the English Language,

CONTAINING,

1. All the words in general use, with their significations accurately explained, and the sound of each syllable clearly expressed; among which will be found several hundred terms, with their acceptions & derivations, which appear to have been hitherto omitted by the best lexicographers; also, a variety of the technical terms of medicine, law, commerce, arts, and general science. The whole interspersed with critical and philological observations, and references to the respective authorities. To which will be prefixed Mr. Walker's principles of English pronunciation.

2. A nomenclature of the names of distinguished persons and places of antiquity; comprising a sketch of the mythology, history, and biography of the ancients, from the most authentic sources.

3. A chronological table of remarkable occurrences from the earliest ages to the present time; containing whatever is worthy of record; as discoveries, inventions, &c. &c.

Compiled from Authors of the most approved Reputation, with considerable Additions:

BY AN AMERICAN GENTLEMAN.

Ad res pulcherrimas erutas alieno labore deducimur.   SENECA.

Hartford Courant, August 14, 1811.

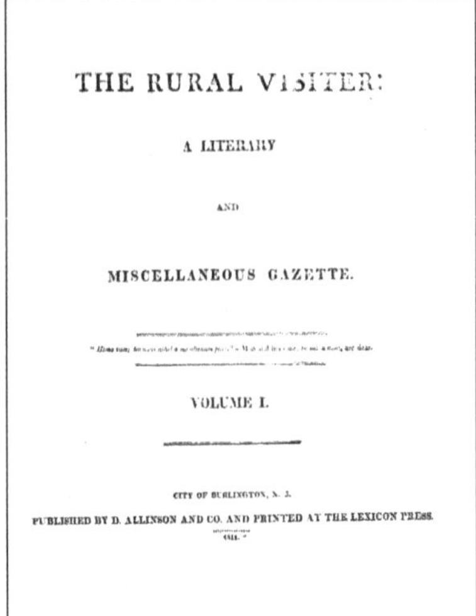

The *Rural Visiter* and the *Saturday Evening Visitor* magazines are available on microfilm at the Burlington County (NJ) Library.

Mary's Protestant Episcopal Church in Burlington but maintained a fondness and interest in Quakerism.

He began a family when he married 31-year-old Philadelphia housekeeper Beulah Zane (1779–1871) at age 50. Although the April 24, 1824, marriage certificate only appears in the general Pennsylvania marriage records,[13] the children's births appear in Quaker records, indicating that the Allinsons raised them in that tradition, despite not holding actual membership in a monthly meeting. A daughter Anne, born in 1825, lived only 3 months. The other children's births include Joseph in 1826, then David Cooper Allinson, born in 1827, and John Cooper Allinson, born in 1828.

In 1810, David owned ten acres of improved land and two horses in Burlington Township, according to the reconstructed federal decennial census. By 1840, the household had grown to include six people. Although no names appear other than David's as head of household, from the ages enumerated, the household included David, Beulah, the three boys, and a young woman who likely served as a domestic servant. Ten years later, in the 1850 federal decennial census, David, Beulah, and their sons Joseph and John remained there, along with several others, all engaged in agriculture.[14]

Near the end of his life, the Burlington Monthly Meeting readmitted David into membership. "In his old age, in much love to the brethren to whom he had long yearned to be reunited, he made a humble and sincere offer to the Burlington Monthly Meeting, which was freely accepted, to his great consolation."[15] Fifteen months later, he passed away. David is buried in the same burying ground with his ancestors. His obituary in the *New Jersey Mirror* on September 23, 1858, reads,

At the residence of his son, near Pemberton, on the 14th instant, DAVID ALLINSON (a member of the religious Society of Friends,) in the 85th year of his age. He was for many years a Printer and Publisher in Burlington.[16] He edited several literary periodicals, the most important of which was the *Rural Visitor*, which extended to three volumes, and was the medium of introducing to the public not a few compositions which "the world will not willingly let die." He published a number of valuable works, legal, literary, and theologic. Among those were a small Dictionary, and a large English and Classical Dictionary. The classical department of the latter is still highly valuable. He never prostituted the press by the publication of profitless trash, such as now floods the world. In the establishment of the Apprentices' Library and in other public efforts, he served his generation in the strength of his days. Of late years he has lived in the country, engaged in Agriculture. Always loving Sacred Truth and approving the things that are excellent, his mind was stored to an unusual degree with the sentiments of the best religious writers—but, as he approached the confines of Time, he realized the sentiment which Scott expressed when dying: "THERE IS BUT ONE BOOK"—and the Bible was his most frequent companion. His aspirations for the highest good were strong—and for this his soul often agonized; and we may reverently trust, that, through the mercy of God in Christ, Jesus, the true end of existence has been attained, and a life of much sorrow and care terminated in complete salvation purchased for him by the Saviour in whom he most surely believed.[17]

David's will detailed specific items left to his widow, sons, his siblings, and some of their children. To his wife Beulah Allinson, he left his clothes, bed, mattress and bedding, tin box and contents, a choice of two or three volumes of his printed books, his shares of stock, and all the money in his bank notes. Joseph was left specific books, as were some of his nieces and nephews; John received his father's desk and its contents (except clothes), all his volumes of *Friends' Review*, garden tools, and pocketbooks, the contents of which were to be used to pay expenses, then the rest to Beulah. Son David Cooper, although named along with John to be executor of his estate, is given only the remaining books that have not already been left to the others, and he is to divide these volumes with John.[18]

Joseph Allison/Allinson
Photograph from family archives.

The Sons

Joseph Allison, the eldest

Joseph, the eldest son, was a talkative man with strong opinions. He was known locally as "The Springfield Orator." To differentiate himself from other family members, he changed his surname's spelling.[19] Much of what we know today about Joseph Allison/Allinson we derived from newspaper accounts and from letters owned by this author's family. Allison traveled and corresponded extensively with family and friends.

During February, March, and April 1855, a series of debates occurred between William M. Ewan and Joseph Allison, who at that time resided in Juliustown, New Jersey. The debates took place in Mount Holly, Pemberton, and Buddtown. Debate topics were: "Does Wealth exert a greater influence over the mind of man than Woman" and "Is War justifiable under any circumstances?" The newspaper assured its readers that "The interest of the subject and the well-known ability of the speakers will draw a crowded house." The February 8, 1855, *New Jersey Mirror* contains this announcement: "Between WILLIAM M. EWAN and JOSEPH ALLISON Mr. Allison takes the affirmative, and Mr. Ewan takes the negative on the question. Admission is 6 cts. Doors open at six a-clock. Discussion to commence at seven. Front row seats reserved for the ladies. February 8, 1855."[20]

The debates were a huge success: "The Lyceum was well filled on Saturday night by persons who were exceedingly anxious to hear the discussion between Messrs. Ewan and Allinson, and all appeared highly delighted with the thrilling eloquence and high-toned argument of the able disputants. The decision of the Judges was in favor of Mr. Allinson, though it is proper to state that the audience was divided on the matter."[21]

Joseph appears to have been a friend of Joe Carr Jr., the editor of *The New Jersey Mirror*. Carr published letters from Joseph in the paper, as well as articles that name him. In addition to his speaking career,

Joseph sold newspaper subscriptions. The 1860 federal decennial census enumerator recorded Allison as a single gentleman residing in Pemberton, with an occupation of "Agent New York Tribune."

A collection of letters that Joseph Allison penned to family members is preserved in this author's family archives. They reveal much about his personality, activities, and travels. His letters contain great details about the weather, time, and money. He lists exact prices and takes pride in saving "mony" [sic]. He also gives precise instructions to his brother John, mother Beulah, and sister-in-law Lucy Allinson, telling them what they are to do for him. When he writes to his mother and Lucy, Allison uses the Quaker second person form of address, *thee* and *thy*. Beulah's background was Quaker, although she may have left by the time of her marriage. These forms of address probably echoed in the household during the boys formative years due to their father's heritage. He does not use this form when writing to his brother John or to his cousin Vashti Shinn.[22]

Allison's letters provide no clue about why he, as a young man from a Burlington County Quaker family, was traveling throughout the country. It may be assumed that both Allison and his well-connected father had extensive contacts due to their writing and speaking. He dispatched the two existing letters below, written in 1853 and 1858, from Baltimore and Camden. They provide no hint about conducting business, or any tourism, nor do they provide clues on why Allison visited a condemned prisoner in Camden. They describe an itinerary that includes Juliustown and Frenchtown, New Jersey; Philadelphia, Pennsylvania; Washington, DC; Petersburg and Alexandria, Virginia; Cincinnati, Ohio; and Camden, New Jersey. In these letters, Joseph spells his surname "Allison," which he does consistently for the rest of his life. These letters have been transcribed just as he wrote them.

> ¼ to 7 Baltimore 8th day April 29th 1853
>
> Dear Brother
>
> I arrived at Baltimore 15 minutes after 7 last evening — We would have got here before that time but one of the iron bolts broke between French Town & Baltimore & therefore we had to stand still & stop 26 minutes which made us 26 minutes later to Baltimore — I am going to Washington to day this morning at 9 o'clk — I am to pay to Washington

from here $1.80 cts — I paid from Philadelphia to this city $2.60 — 1 left Phila at ½ past 12 yesterday P.M. If I had left in the 2 o'clk train I would have had time to have got my ticket before I got on board the cars at Phila. I would only have had to pay $2.50 cts for my ticket but I had not one instant to get a ticket as I walked with all my baggage all the way to the rail road where the cars started for this city — The distance I had to walk thither was at least 2 miles but I saved mony by it — I left Juliustown yesterday at 5 A.M. got Burlington ½ past 7 left Burlington at 8 & arrived in Phila ½ past 9 — I shall go to Petersburg in 2 3 or 4 days I see the cost on some times is $7.50 cts but I can & shall go for $5 or for $5.50 cts.

The first hotel last evening I was directed to with out any thing to eat for a bed only I would have ½ a dollar but at same place they told me only a few doors farther I could get lodging for 12 ½ cts where I went & am now here — I have not had breakfast yet — do not know that I will get any — I hasten on — I desire you immediately to do what Mother proposed & that is get my horse blanket flannel shirt &c from Pages forgot to tell them this as Mother wanted me to but am going to write to Benjamin Antrim or his brother to day or to morrow & I will tell Pages in this way You will be after some of my things — I wrapped up the cushin Mother wanted & if you will take care of my things & do as I ask you to in this letter will give you this cushin & you can have it for Mother or do as you please with it —

I want you to send John Emmons (or get it your self) in the stable of Page to get that new scythe & swath there is there it is good article & I am afraid some body will steal it I hope you will attend to it as next summer I wilt want to cut grass with it if I do not make my fortune with out —

I would be glad if you will go to the farm of young James Shreve (it will be his) & get these apples I bought there — get only the sound ones & sell some of them to pay you for your trouble — When you write if you write before I write again write to Petersburg Virginia-

Keep strict account of the number of bushels of apples you get & pay Benajah Woodward for them & deduct 10 cents as I have paid 10 cts towards them
Love to all your Brother
Joseph Allison

April 29th 1853

P.S. It may be a letter to Petersburg will not get to me but it will not hurt any body if it does not most likely though it will — But when I write again I can let you know where I will be certain — I am certain though I may stay several days in Washington I have just had breakfast & only had to pay 12 h cts for it — On the boat from Frenchtown to Baltimore I had to pay 30 cts for supper — It is the last 50 cts I will pay for one meal I was very hungry or I would have went without.

Camden NJ. May 3/1858
Dear Brother[23]

I left Washington 8 O'clock yesterday & arrived in Philadelphia 5 before 4 which I think was a very quick passage as from Washington to Phila I was only in the cars from 8 to 15 minutes to 10 A.M. coming from W. Baltimore I was only in the cars from 11 o'clk A.M. to 15 minutes before 4 o'clk P.M. from Baltimore to Philadelphia — We were in the cars that carried the United States mail that they go faster I think than other lines generally do — I was from half past 12 o'clk last 5th day going from Phila. To Baltimore — going last 5th day to B. I paid $1.60 cts coming from Baltimore yesterday as I was in the line that goes the most direct route from B. to Phila. I paid $3 — I saw Cousin Samuel[24] in Philadelphia about an hour ago only a few minutes before I crossed the river from Phila to Camden where I now am — We ran against a bull between Washington & Baltimore he was small & therefore I think young that I am afraid he will die from the wounds he received — The cow catcher broke his leg or running against him did this — I saw him before he was disentangled from the cow catcher which is made so as just to murder cows or bulls or any thing else that may get on the rail road track — Martin Buell conducted me to the patent office in Washington where there is much to be seen & where I saw a pair of pants a coat & a vest that were worn by & belonged to George Washington — Sunday after noon I went to Alexandria in Virginia & returned the Sunday to Washington city — fare each way 12 ½ cts distance 8 or 10 miles — I expect to have returned home today & wait till middle of the month to go west to Cincinnati Ohio with J. L. Lamb & T. L. Norcross when they to the latter told me he should go the middle of this month — I want to keep as near all of the mony as I have now got to pay my expenses west — I am told it is much cheaper to travel west than it is to travel south — I had thought of going to Petersburg but have given that out — If you have done nothing with that note of F. Heeler keep it till I come When it goes into Bank I expect it will have to have my name on the back of it — I may be in Camden all this week (if the trial of Phillips & English comes on tomorrow as I am told it will) if the case of English & Phillips takes up that time of the attention of the court — If it was not I expect this case to come on I should come home to day but if it comes on to morrow I shall stay in Camden till it is got through with & brought to a close — was told just now I can get board for $2.25 cts per week in this city (Camden) Write to me immediately let me know if Cooper[25] has been at home — I think you had better sell Coopers horses for any price rather than keep — tell Cooper he may rest assured he now can get more for them in my opinion than he can now — This spring horses XXXXX Surprised if they get through with it in one day as I think it will be only what had been & it will only be to say what has been said that — I presume for this they will cut the matter short — I shall go & see A. Spring this afternoon — have no doubt I can see him — he is to be hung 10th day of next month — I shall come as soon as the Phillips case is disposed of

Truly your Brother
Joseph Allison

P.S. It may not be worth while for you to write as it may be I will not be here but it can do no harm to write & write to Camden It may be I will be in Camden all this week & it may be I will not be there only a day or two — I am now in Phila, My flannel shirt was in my horse blanket hope you have got them ere this — I have had a notion of going to N. York directly from here to see O. Hoffman esq — I go write to Camden & let me know if there is any thing new   To John C. Allison Pemberton N.J.

❧

David Cooper Allinson and two of his bird dogs.
Photograph taken by C. H. Brown, Photographer, 29 ½ East State Street,
Trenton, N. J. printed as "cartes" $1.50 per dozen; from the family archives.

David Cooper Allinson, middle child

The middle son shared his maternal great-grandfather David Cooper's name. He followed in the professional footsteps of his paternal grandfather Samuel Allinson. Given his family's association with the Burlington Academy, it seems reasonable to assume that his early education occurred there, although his obituary in the *Trenton Times* identifies it only as a Quaker school.[26] He left home at age 19 for his higher education, and attended the Williston Seminary in Hampton, Massachusetts. At age 22, he studied at the Amenia Seminary in Ameniaville, New York, in preparation for admittance to Yale University. His health gave way, forcing him to withdraw, and travel for recuperation.

He completed his law school education at a later age. The 1860 federal decennial census enumerates him as a 33-year-old law student in Trenton. In February 1863, Cooper was admitted to the New Jersey Bar. He maintained his practice until the end of his life. In addition to law, Cooper studied medicine, and worked as a pharmacist in both Burlington, New Jersey, and in Massachusetts.

His family called him "Cooper" but professionally he used the name D. Cooper Allinson. He was a well-known Trenton, New Jersey, attorney who specialized in debt collection. His advertisement appeared in almost every issue of the *Trenton Evening Times*.

"Several years ago he gained a reputation for himself as a detective by policies and work which rid Trenton of a number of 'quack' doctors. He learned of the presence of this class of swindlers and started after them. His raid was started with innumerable civil and criminal suits, the outcome of which was total riddance of the band of 'quacks.' "[27]

Cooper's lengthy obituary, published in newspapers throughout the country, described his law office as a "curiosity shop," full of "books

D. COOPER ALLINSON, LAW OF-
fices No. 26 East State St. Collecting bills
a specialty.

Trenton Evening Times, May 26, 1899, page 5.

that were falling apart from age. The side walls were filled with filing racks of his own design. All his office furniture had a history and he religiously avoided new wrinkles in law office fixtures."

Cooper held diverse, unusual interests, often deemed newsworthy. "One of his hobbies was bicycle riding, in which he indulged almost without limits. He owned one of the old Columbia machines, which was equipped with solid rubber tires . . . carried with him on his bicycle all the articles which he considered necessary to comfort. . . . His bicycle uniform was unique in many respects. He was usually seen in a light linen hat and linen duster. He inevitably wore leggings."[28]

His inseparable companion was his Irish Setter dog. He was an avid fisherman who designed and made his own reels.

Even years after his death, he was remembered as "a constant source of merriment to . . . fellow practicioners [sic], and many amusing incidents to be related . . . but time and space do not permit."[29] For example, ". . . the yacht of D. Cooper Allinson, which was made in shape of a fish and was one of the big curiosities on our Delaware River years ago."[30] Cooper contended the commonly used boat construction was "unscientific and opposed to common sense." His own design featured a flat prow and pointed stern.[31]

At some point, Cooper adopted a young cousin, Caroline (Carrie) Foster. No documentation has been found for said adoption, so it may have been an unofficial financial arrangement for supporting the young woman and her mother when Carrie's father died. She was the daughter of Cooper's Uncle Asa Robbins Foster and his wife, called Aunt Harriet, of Springfield, New Jersey. They were related by way of mother Beulah. This couple's name appeared often in letters from Joseph. No evidence exists that Carrie ever lived with Cooper.

Contention rankled the family. Following father David's death in 1858, it appears bad feelings developed between Joseph and Cooper, apparently over their father's estate and its settlement. Older brother Joseph openly criticized Cooper, particularly in situations involving money, referring to Cooper as "that scamp in Trenton." In a November 22, 1862, letter to his mother, Joseph wrote that his brother was not contributing his fair share to expenses relating to settling their father's estate, implying that Cooper had treated the family badly; "Thee knows we have one time been true friends to Cooper & we

David Cooper Allinson. Photograph taken by Pine, 27 and 29 East State Street, Trenton, N. J.; from the family archives.

both know how he had treated us afterwards in regard to my final settlement with the execution. . . ."[32]

In his younger days, Cooper became politically active. As he aged, he made less frequent public appearance, so that the newspapers reported any forays to social events with great interest.[33] Journalists also recounted his legal cases, including one in which his client skipped bail, another in which he represented his nephew who fell behind in his rent remittance, and a high-profile divorce case.

Following the Civil War's end, he practiced law in Trenton. He also held a second position for several years as a federal government mail contractor, earning $3,838 per year by 1908. His duties included transporting and delivering mail along a specific route, mail security, and recordkeeping.[34]

Other than his birth record, no evidence exists that he joined any monthly meeting. He is, however, buried in the Burlington Friends' cemetery with many relatives.

Cooper's will offered a surprise. His law office contained many items that actually belonged to Pauline R. Taylor, his "efficient and faithful clerk" of ten years. He directed that these be returned to her. In addition, ". . . to my Cousin and adopted daughter, Carrie P. Hattersley,[35] . . . my nephews William W. Allinson, Charles E. Allinson and my niece Emily L. Allinson,[36] all of my property, real, personal and mixed . . . share and share alike."[37]

Several newspapers summarized Cooper's colorful life, his death, and his burial. *The Philadelphia Inquirer* stated he was "one of the best known lawyers in the state."[38] The local New Jersey papers provided more details.

> D. Cooper Allinson, known as the most eccentric lawyer in New Jersey, died in the home of relatives in Burlington. He was seventy-nine years old and had lived in Trenton nearly all of his life. He studied medicine, and pharmacy as well as law, and earlier in life had a reputation as a detective. He never married, lived out of doors summer and winter until crippled by rheumatism, when he took to bicycle riding to effect a cure. One of his fads was to try all the new breakfast foods placed on the market.[39]

It was his dying wish that friends should not jeopardized their lives by standing beside his grave bareheaded.[40]

D. Cooper Allinson, who was born near Burlington on April 27, 1829, a man known for his many eccentricities, died on the farm of his nephew, Charles E. Allinson,[41] outside of Burlington, on Thursday. Mr. Allinson was a lawyer, and much of his life was spent in and around Trenton. For many years he lived along the banks of the Delaware river, near Trenton, in a houseboat of his own construction, and spent much of his spare time fishing. He enjoyed outdoor life and in carrying out his ideas he attracted much attention. He often said that "funerals beget funerals," and he made the request that in the event of his own death the mourners should not bare their heads at his grave if inclement weather prevailed. The funeral took place on Monday.[42]

John Cooper Allinson; photograph from the family archives.

John Cooper Allinson, youngest son

John took a very different life path from his father and brothers. Unlike his brothers, he lived his life outside the public eye of the press. He became a farmer. In a December 24, 1861, letter to his cousin, Joseph reported that John and Cooper jointly purchased the Jonathan Fox farm in Pemberton, New Jersey. An 1876 map of Pemberton Township[43] shows that John also owned properties in town. John kept some of Cooper's belongings at the farm, including horses.

John and Lucy Ann Leaver of Shrewsbury were married in a Quaker ceremony. The Burlington Monthly Meeting minutes record that the couple appeared before the meeting to declare their intent to marry; that they had their surviving parents' consent; and that the meeting affirmed the proposed marriage. The minutes then reported that John on ". . . this twenty-third day of the fourth month in the year of our Lord one thousand eight hundred and fifty-six . . . did . . . openly declare that he took . . . Lucy Ann Leaver to be his wife, promising, with Divine assistance to be with her a loving and faithful husband until death shall separate them . . ." Following the Quaker custom, the marriage certificate included the names of all 76 people present, many of them named Allinson.[44]

Lucy was a first-generation American on her English immigrant father's side. Both parents were Quakers, and the Upper Springfield Monthly Meeting minutes record her birth. John and Lucy remained Friends all their lives. They held membership, at different times, in

Upper Springfield Monthly Meeting Record of Members, Births, Deaths.

John's wife, Lucy Ann Leaver Allinson; photograph taken by William Nick, "Excelsior" Traveling Artist, from the family archives.

both the Burlington Monthly Meeting and the Upper Springfield Monthly Meeting, where the births of their children are recorded.[45]

The 1860 federal decennial census shows John as a 29-year-old farmer, living in Pemberton, New Jersey, heading a household that included his wife Lucy, four children, and his 65-year-old mother, enumerated as a "gentle lady." He owned real estate worth $7,000[46] at that time and held a personal estate of $1,500. The household also included an 18-year-old farmhand, and an 8-year-old girl whose familial relation or role is not discernible.

Brother Joseph resided nearby in a Pemberton boarding house. Joseph relied heavily on John and Lucy. He wrote them both often, giving lots of orders for things he wanted them to do.

A Nation Divides—The Brothers' Response

After years of increasing tension, the country was in crisis. The Presidential election of 1860, won by Abraham Lincoln, was contentious to those who opposed Lincoln's unwillingness to compromise on the issue of slavery.[47] His election was followed by the secession of the state of South Carolina and others.

> The State of South Carolina having resumed her separate and equal place among nations, deems it due to herself, to the remaining United States of America, and to the nations of the world, that she should declare the immediate causes which have led to this act. . . . An increasing hostility on the part of the non-slaveholding States to the institution of slavery, has led to a disregard of their obligations, and the laws of the General Government have ceased to effect the objects of the Constitution. (excerpt, *South Carolina Declaration of Secession*, December 24, 1860)

Although New Jersey started down the long, ragged road to eliminate enslavement within the state beginning with the 1804 Gradual Manumission Act, slavery did not come to a complete end until other states ratified the 13th Amendment to the United States Constitution. Always reticent to totally remove the forced servitude stain from New Jersey, the legislators did not vote to ratify the amendment until January 23, 1866, several months after its ratification. New Jerseyans and their elected officials were divided and confused. Most opposed secession and sought unity. Others wanted compromise for a peaceful resolution. Commerce with the agricultural southern states proved essential to the burgeoning industrial north.

The South Carolina Militia crossed the line in the sand when the troops attacked Union forces at Fort Sumter on April 12, 1861, initiating the conflict known in the North as the Civil War. On April

15, 1861, three days after the bombardment, President Abraham Lincoln issued a proclamation calling for 75,000 men to establish a federal militia to serve three months. New Jersey's quota was 8,123 men, which the state met. In fact, the number of volunteers exceeded the number of available slots, forcing many New Jersey men to enlist in New York or Pennsylvania units[48] until New Jersey governor Charles S. Olden successfully received federal permission to establish additional regiments.

War deeply troubled Quakers as it violated their peace testimony. Friends use the word "testimony" in referring to a common set of historically rooted attitudes and modes of living in the world. The Friends' Peace Testimony is a fundamental belief.[49] They stated it clearly in 1660 to Charles II, the king of England:

> We utterly deny all outward wars and strifes, and fighting with outward weapons, for any end or under any pretense whatsoever; this is our testimony to the whole world. . . .[50]

Enslavement proved equally distressing; it was a key element of the secession. Quakers profess ". . . there is the light of God in every person, and thus we believe in human equality before God."[51] Prior to the American Revolution, Quakers had wrestled with the slavery issue, which they viewed as both abhorrent and an economic necessity in some cases. In 1774, the Philadelphia Monthly Meeting reached a consensus that all members should discontinue buying and selling slaves and required the members to prepare for freeing the slaves they owned from bondage at the earliest opportunity. The Burlington Monthly Meeting minutes do not say when the following sale occurred, but contain this entry:

> . . . the 6th day of the 1st month 1817—Samuel Allinson offered in writing a condemnation of his conduct, in having violated our Testimony against Slavery by advancing money for the purchase of a coloured [sic] man who was in bondage and in suffering his name to be inserted in a bill of sale to convey him away with a view of his serving a term of years. . . .[52]

Emerging trends of the Enlightenment, a rise in Protestantism in the United States, and growing liberalism impacted Quaker ideas and practices. A "Great Separation" occurred during 1827, dividing the Religious Society of Friends along ideological and socio-economic lines into Orthodox meetings and Hicksite (named for leader Elias Hicks) meetings. Orthodox meeting members tended to be more financially successful and progressive, placing more emphasis on Biblical authority and atonement. The Hicksites sought to reject what they considered corrupting influences, and to return to the purer, original concepts of Quakerism, with emphasis on the Inward Light. Many communities hosted both Orthodox and Hicksite meeting houses, some of which were built side by side. Approximately two-thirds of Burlington County meetings were Hicksite. The Burlington Monthly Meeting remained Orthodox.

Despite having different religious practice viewpoints, both groups worked towards the abolition of slavery and war and promoted the relief of other social ills. Members of both played a role in the "Underground Railroad," giving aid and shelter to freedom seekers as they fled to northern states or Canada. The Philadelphia Yearly Meeting (to which the Burlington Monthly Meeting belonged) became one of the nation's most prominent pacifist organizations. Quakers were subject to the draft but were permitted to prove why they should be exempt. Fewer than 150 Quakers were ultimately drafted, but many enlisted voluntarily.[53] Monthly meetings appointed investigative committees to determine if members failed in adhering to the sect's testimonies and then fined or disowned (expelled) members not living properly.

So, Quakers were torn. On one side was the Peace Testimony. In addition to refraining from fighting, Quakers should not participate in governments actively engaged in warfare or contribute to promoting war through business or other activity. On the other side was the condemnation of slavery and the goal of abolishing it. Friends faced picking the lesser of two evils, and risked alienation from their congregations either way.[54] Secession put the continuation and westward expansion of slavery front and center. Preservation of the Union created an additional complexity layer. The Testimony of Integrity places God at the center of everyone's life. One soldier wrote about his decision, saying, "I came to the conclusion by serving my country I would be serving my God and friends."[55] As such, integrity is in choosing to follow the Spirit's leading despite the challenges and urges to do otherwise. While opposing war and counseling against participation in it, Quakers "hold in love our members who feel they must undertake it."[56]

Beulah Allinson, a 68-year-old widow living with her youngest child's family on a New Jersey farm, must have felt the weight of this struggle as she watched each of her three sons respond in his own way. They were not young men; all three were in their thirties when they made their choice. Two of them donned a soldier's uniform and marched off to war. Only one of them would arrive home alive. The third son threw his considerable energy into alleviating the pain and suffering the war caused.

Joseph—First born and first to go

Joseph possibly joined the thousands of men who flocked to Trenton and milled about while Governor Olden sought to resolve dealing with too many volunteers. He was, however, certainly there from the beginning. On May 8, 1861, at age 35, Joseph Allison enlisted in the First New Jersey Infantry Regiment, Company D, and on May 22 he was mustered in as a Private for a three-year term.[57] It appears both patriotic fervor as well as pursuing slavery abolition motivated his enlistment. Just five years earlier, he publicly debated the question "Is war justified under any circumstance?" He argued in the affirmative and was judged to be the winner.

While not specifically mentioning slavery, Joseph was mindful of it. For example, in April 1862, he wrote home from Cedar Creek and Catlett Station, Virginia, saying,

> We often hear the cannons roar here it is truly like war but we have come here prepared for the worst we only want to have our camps trimmed & our lights burning we know a day or an hour of virtuous liberty is worth a whole eternity in bondage I would rather therefore die in a good just right cause where we are than live in Confederate states with rebels in their unjust unholy combinations against Liberty Freedom Union & Independence It was this our forefathers fought & bled & some (very many) died for to bequeath & hand down to

their children & their childrens children equal rights equal laws Union & Independence an inheritance uncorruptable undefiled & that fadeth not away.

History books record that the regiment left their Trenton, New Jersey, camp on June 28, headed for Washington, D.C. On July 12, they crossed the Potomac River into Virginia, headed for Centreville, and entered the debacle at Bull Run on July 21. Following that, the regiment returned to camp in northern Virginia. The family archives contain no evidence of Allison's whereabouts at that time. The first letters in the collection are from Camp Seminary, Virginia, in September 1861. No existing letters contain any reference to that battle.

The regiment spent the first fall and winter in Fairfax County, Virginia, near Alexandria. Allison knew this place well, for he had friends here. From his early letters, it appears he greatly enjoyed the army and his encampment, and he believed the rebellion would be put down quickly with little bloodshed. He was often sick. War profiteering upset him, and he described the Union troops looting and inflicting vandalism. He looked forward to receiving his weekly *New Jersey Mirror* newspaper. Newspapers were important to the troops. One writer said, "The papers went like gingerbread at the State Fair!"[58]

He was cocky and sure of himself. After being in service just six months, he wanted a furlough to visit home. He writes, "I intend to have a furlough some time next month & if I cannot get one of our Superior Officers to get me one I will write to President Lincoln or Gen. McClellan & I am sure I must succeed in the end." There is no documentation to indicate whether the President or the General acceded to his request.

Joseph was a prolific letter writer, which should be no surprise, given his upbringing, and he knew a great many people. All the people mentioned in his letters appear in the Index; those who were in the military appear in Appendix A of this book. Joseph wrote some of his letters to his brother John. Others he wrote to the person he addresses as Sister Lucy of Pemberton; however, that was actually Joseph's sister-in-law, John's wife.

A December 4, 1861, letter from Camp Seminary in Virginia illustrates that he still recruited newspaper subscribers even while at war. He also fulfilled the role of a war correspondent, sending first-hand action accounts to his contacts at the *Mirror.*

It appears Allison could leave camp frequently to visit his Northern Virginia acquaintances. He reports home about the fine dinners he enjoyed with friends. His letters contain much information about his ill health. He frequently feels poorly and often required the camp doctors' services. In fact, he insisted on it.

> We had a grand review last Thursday it gave me the head ache & I have not got clear of it yet; I tried to get Doctor Gordon to excuse me for duty today, though he gave me (I mean Dr.) 4 pills & said after it I must take castor oil I took the pills thee sent instead yet Gordon said he had it not in his power I went to Major Hatfield & Mr. H could not do it & said Doctor was only one that could do it.[59]

The early letters are arrogant, full of flowery, bombastic, heroic platitudes. Death and disease prevailed in the camp.[60] Although there had been some casualties, he had not yet seen the horrors of war. "There was a drummer died of the rebels here day before yesterday in company A. he was very pale yesterday when saw him I guess he had the consumption."

Joseph witnessed the enslaved population of the south firsthand. In one letter, the regiment had arrived at ". . . the beautiful & very valuable residence of Doctor Wm F. Gains this man has 13 hundred slaves." In another letter, he comments, "there is a vast difference from what it is in the free states—Truly out south here they have gone for cotton & the "peculiar Institution of Slavery"[61] rather than go for something, even at this time would bloom, blossom & flourish more as the rose."[62]

The tone of the letters changed after the Regiment departed Alexandria. The war caught up with him. Following the disastrous Battle of Gaines Mill on June 27, 1862, in which the 1st Brigade suffered 1,000 of the estimated 15,500 casualties[63] he wrote a letter published in the July 24 *New Jersey Mirror*:

> I was only five or six feet from James Flood when he was shot dead. He was in the same tent with me when we were at Fairfax Sem-

inary. Another man fell dead close to Flood, but he fell on his face, and I do not know whether he was one of our company or not. John Gano, in our company, was about ten feet from me, and never spoke after he was shot We are out night and day on picket[64] or fatigue[65] duty. Night before last, at 7 O'clock, our Company went to work in trenches. We returned to camp at 1 o'clock in the morning. A part of our Company had to go on picket again yesterday, and the other part of us to work at entrenchments.[66]å

He fought in the second Battle of Bull Run at Manassas, Virginia, on August 29–30, 1862. The September 4, 1862, issue of the *New Jersey Mirror* says on page 1, "Joseph Allison of this vicinity, we regret to say, had his leg shot off in the recent battles." Following the battle, Allison was transferred to a Washington, D.C. hospital, where he apparently was not a very cooperative patient. Again, the *Mirror* reported the information in a letter to the editor on September 18, written by E.S.

I had almost forgotten to say I have in my ward Joseph Allinson, known as the Springfield orator. His leg was amputated about two weeks since; he is doing very well; and would do better if he was more obedient; but do as will, coax, threaten, or scold, it is always the same; his knowledge is equal if not superior to my own, and his own way he will have; he is exceedingly loquacious and indulges the hope of getting back to Burlington County and having a bout with the great men.

He never got that bout. Allison died, presumably of infections from his wounds, on October 2, 1862. He was 37 years old. Once more, the *New Jersey Mirror* carried the news. The October 9, 1862, edition says,

Joseph Allinson, of the County, died in one of the hospitals in Washington, on Thursday last (October 2, 1862). His remains were brought to Burlington for interment. He was in all the battles in which Gen. Hooker's 1st New Jersey Brigade participated, several of which were among the hardest fought contests of the war—and though always at his post in the hour of danger, he escaped unhurt till the fight at Manassas, on the 27th of August, when he was wounded—his leg being badly broken by a ball, rendering amputation necessary, from the effects of which he died."

Three days later he was buried in the Broad Street Methodist Church graveyard in Burlington.

John is next

The family archives contain no letters from John. Since he had a wife, mother, and children at home on the farm, he possibly had no need to write to his Shinn cousins. This author consulted public records, military records, and other sources to derive information concerning John.

Days after his brother was wounded at Manassas, John enlisted. On September 1, 1862, the 33-year-old Quaker farmer left his mother, pregnant wife, and four children to join the Twenty-Third New Jersey Infantry, primarily comprising recruits from South Jersey farms. Men from the same county or even the same towns often served in the same company. Most already knew each other; some were related.[67] A cousin, William H. Allinson, entered Twenty-Third's Company A. People initially referred to the unit the Quaker Regiment because many Burlington County residents who served also held membership in the Religious Society of Friends. Although there are no letters from John in the family archives, the accomplishments of the Twenty-Third Regiment are well documented.

John mustered at Beverly's Camp Cadwallader which only one month prior had been renamed from Camp Stratton,[68] adjacent to the railroad tracks. One of his comrades-in-arms recorded, "There are a great many people here to day we expect to get our new suits this afternoon our Regiment was organized yesterday we are the 23rd Regiment of New Jersey Volenteers [sic] Remember me to all of my family and friends you had better bring or send me five dollars for I do not know wether [sic] we will get our first months pay or not. . . ."[69]

Speaking of pay, anxiety grew among the men about their pay schedule, and how to send it to their families. "Governor Olden, who has ever been mindful of the wants and interests of the Jersey boys, has provided for this case by sending Col. Jonathan Cook of Trenton, who says he expects to deliver to our families living in this state, in person."[70] John sent money home. The "Amount of Money Received from Members of Company I of the 23rd Regiment, New Jersey Volunteers by Jonathan Cook, Trenton, New Jersey, to be delivered to . . . free of charge" records indicate that John Allinson allocated $80 to D. Cooper Allinson, Trenton, New Jersey. No date appears on this record page.

Soldiers in the field were supposed to receive pay every two months, but it more likely happened every four months or more because the Paymaster traveled long distances to reach the troops.[71] To receive their pay, troops mustered and paraded by company. Each company commander attended the muster and maintained a required presence at the Pay table.[72] Muster rolls held at the New Jersey State Archives indicate John receiving his pay on October 1, 1862, February 28, 1863, and June 27, 1863. This last payment lists "Amount for Clothing in Kind, or in Money Advanced: $34.08."[73] The cost of his uniform was deducted from his final pay.

Departing camp on September 13, 1862, the Regiment marched approximately a mile to the wharf on the Delaware River, and boarded steamboats to Philadelphia.[74] Cheering throngs lined the street, both in Beverly and Philadelphia. They marched through the city and then boarded cattle cars headed for Washington, D.C. to join the Army of the Potomac. There, the unit combined with other New Jersey units to form the First New Jersey Brigade. The Brigade served to replenish units that had suffered numerous combat casualties and required new troops. At some time during this assignment, John may have received word that his brother Joseph had died in a Washington hospital.

They were initially armed with antiquated flintlock muskets, poorly equipped, poorly trained, and poorly led. A new commander, Colonel Henry O. Ryerson,[75] arrived in November to take over the regiment. He found them a boisterous, undisciplined band of citizen-soldiers. He "dubbed his new command 'The Yahoos,' perhaps not as a compliment. The name stuck and the men of the Twenty-Third emblazoned it on their battle flag. They also proudly called themselves 'Yahoos' until they were in their dotage."[76]

In late November, the First New Jersey Brigade received orders for Fredericksburg. On December 7, Col. Ryerson wrote, "The weather last night and today as cold as any . . . the ground covered with snow—the men with nothing but their shelter tents to protect them from the cold—only green wood and very little of that, scarcely any axes to cut it with—one blanket to a man and many of them shoeless. . . . I almost despair while thinking . . . of this terrible winter campaign."[77]

As the troops crossed the Rappahannock River, the men of the Twenty-Third held the first line as they came under enemy artillery shelling. The real baptism by fire occurred at Fredericksburg, where the Twenty-Third received orders to foil a Rebel counterattack. The effort collapsed and the regiment withdrew at the cost of five men killed, 36 wounded, and nine missing.[78]

Back across the river, the Yahoos established winter quarters near White Oak Church, Virginia. [79] Terrible conditions prevailed in Camp White Oak. The camp possessed no tents and not enough blankets, while the troops contended daily with poor sanitary conditions. Severe winter weather descended on the camp, the worst since George Washington's troops camped at Valley Forge. Many men succumbed to the rampant disease prevalent throughout the camp.[80] Over the winter, Col. Ryerson was reassigned, with the opportunity for promotion. He was replaced by young E. Burd Grubb, a member of the Jersey Brigade staff.

They did not remain encamped for long. On January 20, 1863, the Union Army of the Potomac began an offensive. The intention was a simple, quick dash to establish themselves at the rear of the Confederate army. All went well until torrential rain started falling that evening without letting up. The troops worked all night to advance, but the unrelenting rain continued. The "Mud March" was beyond endurance. "The flow of men, animals, and wagons turned roads into soup. Regiments were forced to stop and rest every 100 yards just from the effort of walking through the mud. Wagons and cannon sank to their axles and were abandoned as not even the efforts of a dozen horses could move them."[81] One war correspondent wrote:

> The ground had gone from bad to worse and now showed such a spectacle as might be presented by the elemental wrecks of another Deluge. An indescribable chaos of pontoons, vehicles, and artillery encumbered all the roads—supply wagons upset by the road-side,

guns stalled in the mud, ammunition-trains mired by the way, and hundreds of horses and mules buried in the liquid muck. The army, in fact, was embargoed: it was no longer a question of how to go forward—it was a question of how to get back.[82]

By morning on the second day, it became clear that the Union forces had lost the element of surprise. Forced to retreat, Allinson and his fellow Jerseymen heard a humiliating chorus from the Confederates, jeering and insulting the Union regiment across the river.

Back in New Jersey, political turmoil ruled the day. Some state legislators, given the dubious moniker "Copperhead" (a venomous snake found throughout the South), advocated a peaceful compromise with the seceded southern states. In support of their troops, Burlington County citizens began a subscription fundraiser to purchase a full-color regimental flag for the Twenty-Third. Many Sunday School children presented the emblem to the troops on April 18, 1863. The presentation ceremony was stirring. According to records, a newly appointed lance sergeant received instructions to protect the flag with "honor . . . dearer than life." As the band played, the Yahoos cheered for the "Sunday School Army of Burlington County."[83]

Within days, the men departed again for Fredericksburg, where they would support other units already militarily engaged. As they advanced, the Twenty-Third troopers became separated from the rest

Salem Church with Veterans visiting after the war. Photo courtesy of the National Park Service.

of the brigade. Coming under heavy fire as they approached the Salem Church, the Twenty-Third advanced towards the woods filled with Rebels behind the church, who then launched an ambush. The Yahoos rallied, and drove the Rebels back, but the ensuing confusion and devastating, withering fire caused the regiment to give way. It was an impromptu battle; the Division Commander did not anticipate a fight nor planned for one. "The losses of our Brigade near the church were severe. . . . Besides the wounded lying thick along our way, prisoners were taken in the woods and in the little gullies in the open field."[84] In a letter the Master Sergeant wrote "We lost 16 killed and wounded in the fight and got badly wiped."[85] Col. E. Burd Grubb wrote to his commanding officer that his men "all behaved nobly." The *New Jersey Mirror* reported the action this way on May 14, 1863:

The 23rd Regiment—which is in the First New Jersey Brigade—was in the recent battle, and lost about 130 men, in killed, wounded and missing. A correspondent writes as follows: We have had two days of hard fighting.—The attack upon the enemy's works, was made on Sunday at daylight, and by nine o'clock, we had carried them and the rebels were in full retreat. We followed the retreating foe to Tabernacle Church, near Morrisville, about three miles south of Fredericksburg, where they had received thirty thousand reinforcements—the division of Gen. Longstreet. Our brigade had the extreme advance, and we charged upon them, but when we met them, we found they greatly outnumbered us—yet we were successful in holding them, until the corps arrived—and soon one of the most hotly contested battles of the war, was fought. Our Brigade was under a terrible fire. Charge after charge was made by the entire corps, but everything seemed to fail. The enemy greatly outnumbered us, and they were strongly entrenched. We fought until it was dark, when the Brigade was ordered to retire. On Monday the battle of Tabernacle Church was renewed, and about 3 o'clock, P. M., the enemy had turned our left flank, which enabled them to take possession of fighting.—Fredericksburg and the heights—thus compelling us to evacuate to the south side of the river.—We started for Banks' Ford, reached it about dark, and long before daylight, we were on the north side.[86]

In addition to the killed and wounded that day, the Twenty-Third lost men taken as Confederate prisoners, including John Cooper Allinson. "We have just learned that a number of the 23rd Regiment, who have been reported as killed and missing, were taken prisoners, have since been paroled and are now at Annapolis."[87]

Neither side could sustain the burdens of sheltering and feeding prisoners. The two armies developed a system of prisoner exchange, which proved successful during the war's early years. When excess prisoners remained after the exchanges occurred, they ended up at the parole camp in Annapolis, Maryland. These parolees embarked on trains to City Point, Virginia. From there, a Flag of Truce boat would take them up Chesapeake Bay to Annapolis.[88] The parole camp featured separate Confederate and Union containment areas. Parolees were required to sign an oath of honor that they would not take up arms or perform soldierly duties until a proper exchange could be arranged. Usually, they were paroled after ten days and returned to their unit. The Muster-Out Roll contains this notation for John C. Allinson: "President's proclamation May 18th 1863—taken prisoner May 3, 1863—paroled but not exchanged, rejoined June 1863" (underlining appears on original record).[89]

A controversy arose over whether a Dr. George C. Brown delivered boxes to all members of regiment in a proper and timely manner. This apparently took place while the men were at the prisoner camp in Acquia Creek, near White Oak Church. John C. Allinson is one of the signers of a February 26, 1863, letter to the "Citizens of Mount Holly and Vicinity" attesting that Brown did everything in his power and could not be blamed for the circumstances of their detention.[90]

Soon after that, the Yahoos received orders to return to Beverly, New Jersey, for mustering out. On June 27, 1863, John Cooper Allinson again entered civilian life. During his war service, his wife Lucy gave birth to a baby girl, also named Lucy, who died at age ten weeks and three days. Baby Lucy's father never knew her. The 1870 Federal Census shows him, once again, as a farmer living with his wife, four young children, and his elderly widowed mother.

On September 29, 1890, John applied for a military pension, stating that he was an invalid. D. Cooper Allinson served as the attorney of record representing him on the application.[91]

In 1906 the state of New Jersey erected a monument—the only one ever erected for a nine-month New Jersey unit—on the Salem Church battlefield, where it stands today.[92] This monument commemorates the Twenty-Third Regiment veterans, who would be remembered for the Battle of Salem Church for their remaining lives. Their obituaries often mentioned the battle.

Years later, an elderly Yahoo veteran would visit a local classroom and talk about the battle. One student present that day remembered only that the veteran kept muttering "shots and shells, shots and shells" over and over.[93] The terrifying memories remained vivid in his mind.

In December 1888, the veterans of the 23rd New Jersey Volunteers met for the purpose of organizing into The Re-union Society of the 23rd Regiment N. J. Volunteers to celebrate the 26th Battle of Fredericksburg anniversary. The Society agreed to meet every year on May 3, the anniversary of the Battle of Salem Church, and did so until 1934, when the two old, frail surviving members answered, "the last call to arms."[94] The last Yahoo died in 1939. John Allinson's name appears in the *History of the Reunion Society of the 23rd Regiment, N.J. Volunteers* book as a private in Company I of the Twenty-Third, but his name does not appear among the society's membership.

Cooper's decision

Like others, Cooper registered for the draft in 1861. His contribution to the war effort, however, was as a non-combatant, a very different circumstance from his two brothers. It is unknown whether he chose this route in adherence to his Quaker testimonies; because his mother had already made such sacrifice with two sons active in the conflict; whether he felt personally unsuited for the rigors of combat, or some other reason. He reportedly declared that "two brothers had gone to the front, and that he would give his talents to his country in the only way open to him."[95] This tells us that Cooper's decision occurred after September 1, 1862. By that time, both his brothers had enlisted. Whatever his motive, he accepted the laborious task of serving as the United States Sanitary Commission (USSC) attorney, the most important civilian relief organization of the war.

The state of military medicine, with startling poor hygiene, when the war began, adds great significance to Cooper's choice. Men who had never strayed far from home were suddenly thrust into large groups of strangers carrying diseases to which they had never developed immunities. Furthermore, the Union forces encamped and fought in the South's stifling heat and humidity they had never experienced before.

As the war began, Union military leadership failed to develop plans for treating the sick and wounded in the field. Because Union commanders believed the war would be short, they saw no need for any long-standing medical corps. It took Gen. George McClellan, following the first Bull Run disaster, to establish a requirement that one surgeon and one assistant surgeon be assigned to each unit before deployment.[96] No plans or budget

U. S. SANITARY COMMISSION.

ARMY AND NAVY CLAIM AGENCY.

No Charge for Services.

ACTON C. HARTSHORNE,
LOCAL AGENT,
COUNTY CLERK'S OFFICE, FREEHOLD, N. J.

THE U. S. SANITARY COMMISSION desiring to relieve Soldiers, Sailors, and their families from the heavy expenses usually paid for the prosecution of such claims, have established this Agency, to collect pensions, arrears of pay, bounty and other claims against the Government, WITHOUT CHARGE OR EXPENSE OF ANY KIND WHATEVER TO THE CLAIMANT.

On application sent to this Agency, stating the name and post office address of the claimant, the name, rank, company, regiment, service, and State of the soldier on whose account the claim is made, date of discharge or death, the proper blanks will be filled out as far as possible and forwarded to the person applying. These can then be executed and returned to this office, where the claim will be prosecuted to a final issue in the shortest possible time.
Jul15-6m

Monmouth Democrat, July 20, 1865, p. 3. this may be representative of the detailed work David performed on behalf of those fighting for the Union.

existed for transporting the wounded, providing medicine and medical supplies, food, or hospitals.

Private relief and support efforts to aid the troops initially began with thousands of ladies' aid societies throughout both the north and south. They sewed uniforms, raised funds, collected needed supplies, and volunteered as nurses. In June 1861, the United States Congress passed legislation to create the United States Sanitary Commission, a self-funded, volunteer, civilian agency to provide myriad assistance services to U.S. soldiers. The USSC sought to organize the women's groups and their efforts to achieve success on a larger scale. This model had proven itself during the Crimean War with Florence Nightengale.[97]

> The Sanitary Commission remained indispensable. . . . It raised not less than twenty-five millions in money, goods, and personal help. . . . In a nation which had no medical association, no nursing schools, no apparatus for meeting a sudden strain on hospital facilities, (the USSC) mobilized the best talents available for the war emergency."[98]
>
> "From its inception, and for its time, the Sanitary Commission demonstrated a rare fusion of men's and women's energies, equally valued, coming together in common cause.[99]

The USSC promoted clean, healthy camp conditions, raised money for supplies, sent surgeons and nurses to the front, provided battlefield relief, educated the military and government on health issues, investigated troop health condition, collected statistics, and other supplemental services that proved important as the war progressed.[100]

Under the U.S. Sanitary Commission echelon, three departments operated: the Department of Preventive Service, sometimes called the Department of Inspection; the Department of General Relief; and the Department of Special Relief. It is possible Cooper worked within the purview of the Philadelphia office, which managed a pension agency, war claims office, system of special relief for soldiers in transit, and military hospitals. He had received training in medicine and pharmacy, both useful professions while at military hospitals, in providing temporary services for soldiers in transit, or inspections. As a newly minted attorney with a reputation as a detective, his skills would be valuable in many ways. The numerous important supplementary services he performed included aiding soldiers and their survivors to receive back pay and pensions they earned. Papers often were lost in battle. There was also "bewilderment of sickness." Sometimes, through supervisory ignorance or due to the soldiers themselves, claims failed to be filed. One office alone reportedly held over $90,000 in back pay that, without the aid of the Sanitary Commission, would be lost forever or delayed until the soldier's family no longer needed the money. The Commission acknowledged that this service was likely to increase as the war continued.[101] In fact, Cooper represented his brother John in a claim for a military pension many years after the end of the war.

Cooper distinguished himself in his "immense work in collecting for soldiers and their families their pay from the United States. . . .His loyalty to the Union prompted him to do this work without charge. . . . he remained at home and contributed to the Union's cause by repeatedly attacking the disloyal element of Trenton, which element he said 'kept stabbing our soldiers in the back.' "[102] Counsellor Allinson was remembered as "one of the staunchest New Jersey patriots at the time of the Civil War."[103]

The Women Left Behind

Beulah had spent half her life residing in Burlington Township, the wife of a well-known man of letters; and then he died. It was the usual practice for widows to join the household of a married child. There was no question with whom she would live: only one son was married. Of the two quirky bachelor sons, one resided on a houseboat, and the other was a boarder who traveled frequently. Beulah moved in with John and Lucy on their Pemberton farm. Pemberton is far from the city of Burlington, to which Beaulah had ties. What did she think about this life change? Was the countryside too quiet? Did she help with farm chores? Did she miss the social life? Did she make new friends? Was she consumed with taking care of grandchildren? How was she grieving her late husband? Now in her sixties, how was her health? Did she attend Friends' Meeting with her daughter-in-law? Was there a church more like her own available to her? Lucy's own parents lived elsewhere in the county with another daughter.

It is clear from his letters that brother-in-law Joseph depended on Lucy to handle many aspects of his business while he was away. She corresponded with him faithfully; there are many references to having received a letter from her. In one of Joseph's letters, he assigned "Sister" Lucy the task of directing her husband John to collect newspaper subscription fees in Joseph's absence. Joseph filled his letters to her with complaints about what she had done wrong and ordered her to carry out various tasks. Her replies are lost to history.

> I received thy letter & bundle through Ben Nutt Abram says it was not the kind of shirt he wanted I thought thee knew I complained to thee that the government give us plenty of these under shirts but what we want is an over shirt to go over the under shirts we get from the government[104]

> . . . it appears it only takes 6 months or more by thy letter for me to be informed of Mother getting my key I have not been informed before of it yet thee thinks I have forgotten it I have no such a short memory about such matters because I have for more than 6 months kept my eyes & ears open to learn where the key is & has been for a long time — My mother has the letters to me & thee can see whether any one tells if Mother had got my key I am very much mistaken if any one of thy letters give said news.[105]

It is unknown when or how Beulah and Lucy learned about Joseph's wounds, and of his death. Both were reported in the *New Jersey Mirror*. Did someone who had witnessed Joseph's wounds and death visit them to report before they read the paper? It is heartbreaking to imagine anxious families scouring the news every day for any word of loved ones. Was Beulah there when her son's body arrived in Burlington? Was there a funeral? Did she stand at the graveside alone? Did she ever receive Joseph's personal belongings? Did any of his many acquaintances visit to comfort her?

In November 1863, Beulah applied for the pension Joseph had described exactly two years earlier. Did Cooper help her through the process?

> . . . as a single man I can get $24 state pay but I could not get it
> till I returned from the wars — If I get killed as a matter of course I
> would never get it I see a soldier with a wife or having a widow Moth-
> er can draw $5 a month so that will make a difference of $12. . . .[106]

What must Lucy have felt in her heart as her husband departed to engage in a war for a nation to which her own father had so
recently immigrated? What fear the two women must have felt for
John, knowing Joseph's fate. Lucy was three months pregnant when
John enlisted. In addition to running the farm, other responsibilities
included caring for her grieving mother-in-law, and four children
under age five. The summer and fall harvest season yielded to the
harshest winter in years. In late December, with her husband away
in the war, Lucy gave birth to a baby girl. She named her Lucy. Joy
turned to sorrow when the infant died at age ten weeks and three
days. The *New Jersey Mirror* carried the sad obituary.[107] Now, Beulah
and Lucy shared the painful bond of each having lost a baby in the
first weeks of life.

John's homecoming in June 1863 must have been greeted with
great relief, but a soldier who has been through the horror of combat,
deprivation, suffering, and imprisonment, returns a changed man.
Lucy had changed, too. During her husband's deployment, she held
down the fort. She served as head of the household, making all the
decisions, being both father and mother to her children, comforting
her mother-in-law. She buried her dead baby. What adjustments did
both John and Lucy have to make when they reunited?

Beulah passed away in 1871. The obituary appeared in the *Monmouth Democrat*,[108] referring to her as "relict of David Allinson."
Relict—from "relictus"—a Latin term meaning having inherited, been
bequeathed, or being a survivor. Old newspapers used this term as a
more delicate way of saying "widow."

John and Lucy eventually left the farm for a small row house at
225 Penn Street in Burlington, near where his immigrant great-grand-

On the 24th ult., at the residence of her son, John C.
Allinson, of Pemberton, BEULAH ALLINSON, relict of David
Allinson, in the 77th year of her age.

father Joseph Allinson settled in 1713. At age 73, Lucy had her foot amputated due to a bone infection.[109] She died the following month in 1897.[110] John continued to live in the house with three of their grown children until his death on October 3, 1900. Lucy and John are buried just blocks away, at the Friends' Burying Ground behind the Burlington Quaker Meetinghouse. They are surrounded by their ancestors.[111]

PART 2

Private Joseph Allison's Letters from the Front

War is best described by those who have experienced it. For more than a century and a half, this author's family has treasured a collection of letters that relatives on the front lines wrote to those at home. What follows is a careful transcription of Joseph Allison's letters, along with narrative to place the letters in context.

On April 15, 1861, three days after the South Carolina Militia bombarded Fort Sumter, President Abraham Lincoln issued a proclamation calling for 75,000 men to establish a three-month militia force. New Jersey met its quota of 8,123 men. In fact, the number of volunteers exceeded the available slots, forcing many New Jersey men to enlist in New York or Pennsylvania units. Other calls for troops followed. During the war, the state of New Jersey mustered the following organizations:

Cavalry—for three years' service and over, 1 regiment; for three years' service, 2 regiments; total, 3 regiments.

Light Artillery—for three years' service and over, 2 batteries; for three years' service, 3 batteries; total, 5 batteries.

Infantry—for three years' service and over, 5 regiments; for three years' service, 13 regiments and 4 companies; for one year's service, 4 regiments; for nine months' service, 11 regiments; for one hundred days' service, 1 regiment; for three months' service, 4 regiments; 38 regiments and 4 companies. Total—41 regiments, 4 companies, and 5 batteries.[112]

Each regiment had ten companies of about 100 men. Each company was commanded by a Captain. The men were divided into two platoons. The captain commanded the 1st Platoon, and the 1st Lieutenant commanded the 2nd. Platoons were divided into two sections, each of which was led by a sergeant. Each section had two squads, and each of these was supervised by a corporal. . . . Volunteer companies were permitted to elect their officers and non-commissioned officers."[113]

1st Regiment, New Jersey Infantry

The 1st Regiment, New Jersey Infantry, initially assembled at Camp Olden[114] in Trenton, which served as New Jersey's first Civil War soldier training camp. The volunteer troops mustered in on May 21, 1861. The regiment left the State June 28th, 1861, with a full complement of 1,034 men: 38 officers, and 996 non-commissioned officers and privates. The privates included Joseph Allison who had joined on May 8.

The New Jersey Brigade was originally a state militia, with the troops electing the officers, and the governor then appointing those elected. These officers unfortunately had no military knowledge or experience, nor were many equipped to be leaders. The 1st Regiment and the others were poorly trained, undisciplined, and equipped with obsolete rifles. By October, the Regiment had been federalized. Existing officers received orders to go before a Board of Examination to determine their fitness for the job. Most resigned, rather than have their military shortcomings made public.

The Regiment departed for Washington, D.C. by way of Philadelphia. Upon arrival, they became part of the defenses around the nation's capital. The Regiment, attached to 2nd Brigade, Runyon's Reserve Division, McDowell's Army of Northeast Virginia, served in that capacity until August 1861. Allison's letters begin in September 1861, when they joined Kearny's Brigade, Division of the Potomac, until October 1861. Then, orders assigned the regiment to Kearny's Brigade, Franklin's Division, Army of the Potomac, until March,1862. Allison mentions Kearny and Franklin in his letters.

The 1st New Jersey Regiment continued its duty, defending Washington, D.C. until March 1862. In March, they began moving south as part of the Peninsula Campaign. They advanced from Alexandria to Bristoe Station April 7–11, then on to the Peninsula on April 17. Allison dispatched a letter on April 29, 1862, from York County, Virginia, and one day later, he wrote from Cedar Creek and Catlett Station. The siege of Yorktown occurred April 19 through the beginning of May, and a battle at West Point on May 7–8. Allison writes about these in his final letter contained in the collection, dated May 8, 1862.

Fighting continued with the Battle of Gaines Mill on June 27, Charles City Cross Roads and Glendale on June 30, and Malvern Hill on July 1. A *New Jersey Mirror* account of the bloody fighting and its toll on local boys appears later in this book. The Regiment served on duty at Harrison Landing until August 16, moved to Fortress Monroe and on to Manassas August 16–26, and Pope's Campaign in Northern Virginia, August 26–September 2. Allison's final battle occurred at the second Bull Run on August 27, where he "had his leg shot off" and then transferred to a hospital in Washington, D.C.

The Regiment went on to Richmond, just as Allison noted. It remained in existence until the survivors mustered out in Trenton on June 28, 1865. During its service, the Regiment lost 244 men: 9 officers and 144 enlisted men killed and mortally wounded, and 1 officer and 90 enlisted men by disease.[115] Allison died of wound complications on October 2, 1862, apparently having ignored his doctors' orders.

Near Theological Seminary[116]   September 5, 1861
Fairfax Co Virginia
Dearest Cousins,[117]

I am seated in my tent on my blanket it is raining quite hard & it keeps us with in such doors — We are now 9 miles from Washington city 3 miles from Alexandria & 2 miles from Baileys Cross Roads — t have been sick 2 weeks & again assumed the duties & position of a soldier last Friday — We & other companies here every day this week have been at work building a Fort in fact there are 2 here very near us & each company only has to work 3 hours each day — Gen. McClellan was here to see us a week ago. I am in Comp. D there were 10 companies of us in 1st N.J. Regiment 7 6 companies were separated from the three N.J. Regiments for a light battalion I think it is called they have the Springfield Minnie Rifles in the place of the Springfield muskets that we got at Trenton — our Comp. D. & Comp. C. were taken out of 1st N.J. Regiment & Cap, Joe Rowand with his Burlington Comp. was one of the companies that was taken out of N.J. 3rd Regiment we got our rifles[118] & now there are only 6 companies of us together — Col. Montgomery was at the head of our Regiment but Mr. Tucker is our head Col. Now — I just finished a letter to go Barclay White[119] — It rained very steady without any stop — We had a very warm night, night before last I slept with out my overcoat & I have not done for 2 or 3 weeks I think — Last night was not cold but before the last 2 nights we have had cold nights — Give my love to Asa & Harriet Foster[120] We got 6 new spades 6 new axes (6 large & 6 small) some new shoes & last Monday exclusively for our company — I must hasten to close —

The small axes are to use with one hand to make pins to drive in tents — We are building 2 Forts here very near us & I learn that one with in gun shot of us will command 60 guns There are from 300 to 500 men at work all the time at each of these Forts — I did my first work at one of the Forts yesterday our comp. has been at work every day this week excepting today & the rain today has prevented —There is a great abundance of melons & peaches here & a great many pies cakes are brought to our encampment to sell

Love to all I am very affectionately
Your cousin
Joseph Allison
9 Mo. 5, 1861
Write and give me news my address is "Fort Taylor Virginia 1st N.J. Regiment Comp. Care Cap. Mutchler. I am in Comp. D there were 10 companies of us in 15t NJ. Regiment" Joseph Allison

"We are building 2 Forts here very near us & I learn that one with in gun shot of us will command 60 guns — There are from 300 to 500 men at work all the time at each of these Forts — I did my first work at one of the Forts yesterday our comp. has been at work every day this week"
— Joseph Allison

Washington, D.C., located between the Confederate state of Virginia and the border slave state of Maryland, lay vulnerable and undefended. Work began quickly on building a ring of fortifications around the nation's capital. By the end of 1863, 60 forts and 93 batteries surrounded the city, with more than 800 guns controlling all approaches.[121] Fort Ward stood a short distance from the Virginia

Theological Seminary where Allison encamped. Today, the City of Alexandria owns the fort and has restored it as a historical and educational site. It is the best preserved of the system of forts built to protect Washington, D.C. during the war. Based on Allison's location description, and the map depicting the Defenses of Washington, it seems likely that he helped build this fort.

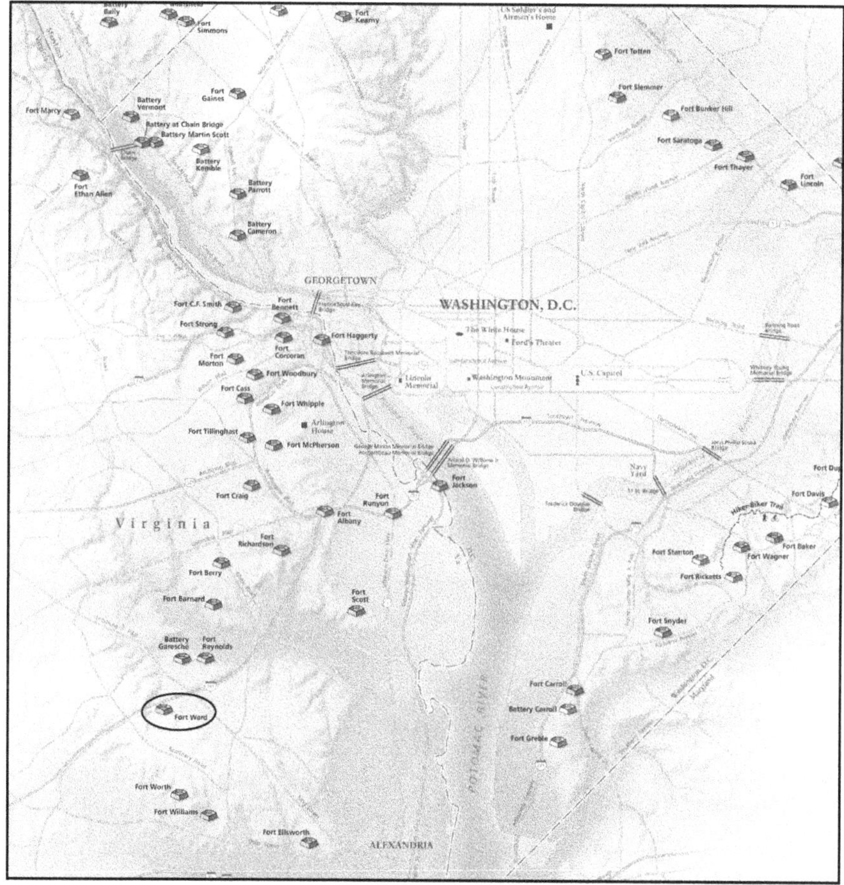

"Construction of Fort Ward began in July 1861, immediately after the Union Army's defeat at the Battle of First Bull Run (First Manassas). The fort was completed in September 1861 and was named for Commander James H. Ward, the first Union naval officer to die in the Civil War. The initial earthwork fort had a perimeter of 540 yards and emplacements for 24 guns. Throughout the Civil War Fort Ward served as a deterrent and never came under Confederate attack. It was abandoned in December 1865, and salvageable materials were sold at auction."[122]

Sunday September 8 1861

Dear Cousin

I received your acceptable letter last evening & at once comply with your request to write soon — It has been a long time since I saw you or was at your house & I was very glad to hear from you as well as the rest of you — I have been intending for a long time to write to some of you but I have been sick ¾ of the time with in the last 3 weeks I thought I had got well last Friday week & I again assumed the duties & position of a soldier on that day but I was taken with again last Thursday evening & got worse & worse & Doctor Welling said I had better come to the hospital Friday morning which I did — Here I am better cared for than I would be at my tent — When I was sick 2 weeks ago I remained in my tent & never entered the hospital on the sick list till last Friday — Here the sick have the very best accommodations as there are several large buildings here & where our hospital is it is in a beautiful large brick building 3 or 4 stories high with a fine large steeple & bell & cross upon it — I was on the top of this Seminary yesterday & it was a most beautiful sight we can see Washington city & Alexandria quite plain from here — I think we can see as much as 20 miles or more from here — Washington is 9 miles from here Alexandria is 3 miles & Baileys cross roads are 2 miles from my camp; it is about a mile from here on the road to Baileys cross roads. Near or quite all of the N.J. Regiments are from one quarter to one mile from the Seminary here — There are two Forts that our companies are at work at by turns each company works 3 hours each day they get a drink of whiskey while they are work & it is fun to see them take it. Some work it so that they get two drinks — I feel much better to day I had an appetite yesterday I am weak but I think I can get out of hospital in two days though I may not as Doctor said I had better stay till I get strong & well — Our Doctor Welling is from Pennington he went to school there when I did & he is an old acquaintance, his Father & Mother lives there — Those of us that can go to dining room got fried salt meat bread & tea for breakfast yesterday soup & bread & tea for dinner & fried beefs liver bread & tea for supper — This morning we got fresh beef in lieu of salt meat — Cousin you set your brothers & sisters a good example to write as you say a beautiful & smart writer truly said "Education is a

Vashti Burtis Shinn (1842–1903) received most of Allison's letters in this collection. She was his first cousin, the eldest of the eleven children of Earl and Emma Eliza Arey Shinn. She was 21 years old when the Civil War began. At that time the family lived in Jacksonville New Jersey. Vashti also received and saved letters from her brother, William Elwood Shinn, who served on the ship U.S.S. James Adger in the South Atlantic Blockading Squadron. Another brother, Joseph Arey Shinn, enlisted in the 3rd Regiment, New Jersey Volunteers. Photographed by William Swaim, Mill Street, next to Farmers' Bank, Mount Holly, N.J.

companion which no misfortune can depress, no crime can destroy, no enemy can alienate or enslave; at home a friend, abroad an introduction, in solitude a solace, and in society an ornament Those here that can read are much more happy than the ignorant that can't read or write"[123] — I hope you will write to me often & I will answer you & give you some war news — We are in our camp only 3 miles from the fortifications of the rebels — There was a drummer[124] died of the rebels here day before yesterday in company A. he was very pale yesterday when saw him I guess he had the consumption — There is 2 months pay coming to us up to the 1st of this month it is said we are to get it next Tuesday then we will get $23 of $25 each I do not know exactly which — We have had a great many different camps since we have volunteered we had camp Perrine & Camp Olden at Trenton camp Monmouth in the District of Columbia near Washington then we come to Virginia & camped at Roaches Mills one mile & a half from Alexandria & on the "Alexandria Lowdon & Hampshire" rail road leading to Vienna from Alexandria — I am with love to you all Affectionately Your cousin Joseph Allison

Envelope addressed to Miss Vashti B. Shinn.[125] Jacksonville, Burlington Co. N.J., Care of Earl Shinn

"There are two Forts that our companies are at work at by turns each company works 3 hours each day they get a drink of whiskey[126] while they are work & it is fun to see them take it. Some work it so that they get two drinks" — Joseph Allison

Constructing a ring of forts to defend Washington, D.C. was so urgent that Gen. Kearny had 12,000 men, working in two shifts, seven days a week accomplishing the task. Often, they received two whiskey rations "to make them work well."[127] As a Quaker, Allison refrained from alcohol. The minutes from the June 4, 1857, Burlington Monthly Meeting summarize their concern over "ardent spirits," agreeing to:

"... appoint a committee to inquire whether any of our members use or furnish that pernicious article as a drink, and if any should be found in that dangerous practice, to endeavor to prevail upon

Aspinwall Hall, Fairfax Seminar, photographed during the Civil War. Courtesy of the Library of Congress.

them to discontinue it. . . .and report to the meeting . . . the number of delinquents."

"where our hospital is it is in a beautiful large brick building 3 or 4 stories high with a fine large steeple & bell & cross upon it" — Joseph Allison

Aspinwall Hall still stands at the Virginia Theological Seminary, located at 3737 Seminary Road, Alexandria, Virginia. It was dedicated in 1859, and still new when it served the Union Army as a military hospital. This is where Allison experienced good care.

"we come to Virginia & camped at Roaches Mills one mile & a half from Alexandria & on the Alexandria Lowdon & Hampshire rail road leading to Vienna from Alexandria" — Joseph Allison

Confederate forces inflicted damage on the Alexandria, Lowdon & Hampshire Railroad three months before Allison wrote this letter. By then, the United States Military Railroad took control of what remained. The garrisons defending Washington received their supplies on this line, which passed through the ring of forts defending the nation's capital, including Fairfax Court House, Occoquan,

Vienna, and Centreville. Vienna was the most important stop and heavily fortified. There, the supplies were transferred to wagons. One source states, "(in) Vienna . . . during the War. A visitor would have seen not so much a town, but a fortress. A stockade surrounded the camp, which occupied both sides of the tracks west of the station."[128]

∾

Fairfax Theological Seminary
Sunday sep 8th 1861
Dear Cousin,

Give my love to your Uncle Asa & Aunt Harriet Foster — their letters are very acceptable — I have a note of it & I think it was the 12th of July we come to Virginia & first camped as I said at Roaches Mills we then went to the residence of Gen. Lee with our tents near Fort Albany — our camp at Roaches Mills we called Camp Trenton- We were only one day at Gen. Lee's residence then we had our tents removed to Camp Princeton near Fort Runyon & very near the long bridge[129] crossing the Potomac into Washington — Fort Albany is only one mile from Fort Runyon — We removed from Camp Princeton to near Clouds Mill to where this Theological Seminary is then we removed from here to where they are building a Fort — They commenced the Fort last Friday week & our tents were exactly where the Fort now is so our tents last thursday week had to be again removed a few hundred yards this way for the Fort — Our tents & encampment are there now but we have been expecting to leave there every day for 2 or 3 days — It was reported last Thursday that the rebels were going to march into Maryland over Potomac at Chain Bridge or at some other point — It was said a very large number were coming & Gen. Kearney give us orders to be ready at a moments warning to march & meet them — But the rebels are such cowards I do not anticipate any thing of the kind — Some of our troops have been continually looking for them to march upon us it is what we want & they would do it to their sorrow — It is said Jeff. Davis is dead & I am sure let it be true or not think it is an ill omen for them — I wrote to Mother yesterday but it was a very short letter as I was not as well as I am to day — but I must hasten to a close — I hope you will write again soon — We have had no name to any of our encampments since we left Camp Princeton at Fort Runyon — We have had a great abundance of melons peaches pies &c here — There is a black woman that makes apple dumplings & brings them to the Seminary building here but she does not take them to our camp — When our encampment was here before we left I got some of her & I got 3 of her yesterday she charges 4 cts apiece with molasses on them — Small pies are 7 cts & large ones they charge 15 cts for; round potatoes here among farmers are $1 a bushel at Alexandria & the prices of market they are one dollar & twenty five cents — In Mirror at Mount Holly i see they are only 40 cts bushel — Last butter I see bought here was 25 cts or 30 cts a pound — Melons from 5 cts to 20 Cts a piece — Stratton & John T. Nixon at Washington franked some envelopes for me when you or cousin Asa Foster write to me tell me if Post Master lets them pass with out pay I presume no difficulty of the kind but my friend Carhart in my tent sent letters to Easton Pa. with Strattons name on & P. Master at Easton charged postage & Carhart has written Stratton about it at Mount Holly making a complaint about it — I got a peach pie made last Thursday at a place where I bought 2 or 3 good suppers I have not touched it yet but I am going to eat a piece this afternoon — I know it is a good one I got one before — Most of those that bring pies cakes melons beer & such things are black men — The most of them live in Alexandria. My address is "Fairfax Co. Virginia, 1st N.J. Regiment Comp D Care Cap. Mutchler." We got Springfield muskets at Trenton — Our comp. now has Springfield Minnie Rifles[130] they were exchanged while we were here in camp — I got a letter from your Father & Mother only a few days ago — I am now going to taste that pie — Write & tell your XXXX

"Stratton & John T. Nixon at Washington franked some envelopes for me when you or cousin Asa Foster write to me tell me if Post Master lets them pass with out pay I presume no difficulty of the kind but my friend Carhart in my tent sent letters to Easton Pa. with Strattons name on & P. Master at Easton charged postage & Carhart has written Stratton about it at Mount Holly making a complaint about it" — Joseph Allison

The United States Post Office proved remarkably efficient during the war, often setting up post offices near encampments. Despite this, soldiers often encountered difficulty obtaining stamps. Beginning in July 1861, the Office allowed Union soldiers to send mail without stamps by writing "Soldier's Mail" on the envelope. Postage was then collected from the recipient. It appears that Allison had a local friend stamp pre-paid franked postage on several envelopes.

Envelope franked for Allison by the local Postmaster.

"We were only one day at Gen Lee's residence then we had our tents removed to Camp Princeton near Fort Runyon & very near the long bridge crossing the Potomac into Washington" — Joseph Allison

Theological Seminary Virginia Sep. 11th 1861
Dear Cousin

When I wrote to you I intended to send you a piece of poetry if you are fond of poetry you will find it interesting & beautiful — John D. Thompson left here yesterday at noon to go to Mount Holly I got 3 letters the day after I got yours — one from my sister Lucy one from Watson Antrim & the other from N. P. Hargrove — I forgot whether told you our camp is only 3 miles from the fortifications & pickets of the secessionists — i was not so well yesterday as usual I had a chill towards night yesterday but it was a light one I hope & think I will have no more I took a powder the Doctor gave me & I think it was that that caused me to be so chilly — Our XXX I was going to send this poetry to you in my other letter but I sealed it up

Sketch of Camp Princeton, ca. 1861. Headquarters of General Runyon's New Jersey Brigade, Arlington, Virginia. Courtesy of Rutgers University, Archibald S. Alexander Library, Special Collections and University Archives.

when it was too late so I thought I could write again & it would be better than to tear envelope to open it.

Letter to Vashti B. Shinn

Jacksonville, Burlington Co. N. J.

From Jos. Allison

". . . our camp is only 3 miles from the fortifications & pickets of the secessionists" — Joseph Allison

"Picket duty was not just guard duty. The picket line was an early warning system. It was actually the front line of the Army . . . and was manned with enough force to delay an attacker until units in the encampments could get into defensive formations."[131]

❧

Theological Seminary Va   sep 23 1861

Dear Cousin

I have very little time to day to write but I thought I must write to you any how & answer your letter which I was glad to get it gave

The Picket at Rappahannock River. Drawing by artist Edwin Forbes (1839–1895). Courtesy of the Library of Congress.

me news that I had not heard — Whiskey has been a great injury to him I think & it should be a warning to some others in Burlington that drink as he did - The same evening I got your letter I got one from J. D. Leaver & he informed me of the death of Judge Swain at Pemberton — I was at Alexandria last Saturday with Thomas Airy he wanted some clothes & I proposed to Mr. Airy for us to go together — We were several times by the house where Col Ellsworth was shot the house was closed Saturday but a beautiful flag was flying at the top of it where Ellsworth took the secession flag from — In coming from Alexandria here we come right by Fort Ellsworth — When we camped near Clouds Mill we come from Camp Princeton very near Fort Runyon & we come direct through Alexandria that was the first time I have been there since my visit to Washington City in the spring of 1853 — In going to said Mill we had about 2 hours in Alexandria & I went to the house & very spot where Ellsworth was shot to see it the house then was open & was in possession of United States troops — Ask your Mother if she remembers lame Henry Clevenger that lived on the Swem property next to my brick house he now lives between here & Alexandria about half way on the road from here to Alexandria — He has lived in this state 3 years — He went from Jersey in 1845 to Philadelphia & lived 13 years then come here — He is just as lame as he used to be — We have a very beautiful clear day to day I have no news to communicate —

The Doctor says this morning I may leave the sick list to morrow — Our company D went out on picket to Baileys Cross roads last Thursday & did not return till yesterday — I am glad to hear from you & young persons can improve themselves much more in writing letters to their friends relations & sweethearts than in practicing at school the reason is obvious the former is done as a pleasure & when we get letters in return we have some remuneration & reward for writing letters but at school it is done as a task & to me it was always a dry one — I never learned to write at school what little Penmanship I am master of I acquired in writing letters & copying select pieces for my scrap book then it was not done as a task or duty — Give my love to all your relations at home & abroad your Uncle Asa has not written to me for some time — I must write a letter to Mother to day or to morrow — I was at Alexandria last Wednesday as well as last Saturday that is 4 times have been at that large place — It is situated on the broad beautiful Potomac — I do not have or take time to go over my letters none of them to correct mistakes or make additions or improvements — It is just 5 months yesterday that I signed the muster roll of Joe Gale at Mount Holly so you see how long it has been since I have had to take the position of a soldier — Affectionately & truly
Your cousin
Joseph Allison

"We were several times by the house where Col Ellsworth was shot the house was closed Saturday, but a beautiful flag was flying at the top of it where Ellsworth took the secession flag from — In coming from Alexandria here we come right by Fort Ellsworth"
— Joseph Allison

Twenty-four-old Col. Elmer Ellsworth, a close friend of President Lincoln, was the first Union officer to die in the war. He became a martyr and symbol of patriotism for the Union.

The day after Virginia seceded from the Union (May 24, 1861), Ellsworth led a contingent of his "Fire Zouaves" to capture the tele-

Theological Seminary Virginia
Sep. 11th 1861

Dear Cousin

When I wrote to you I intended to send you a piece of poetry if you are fond of poetry you will find it interesting & beautiful – John S. Thompson left here yesterday at noon to go to Mount Holly I got 3 letters the day after I got yours – one from my sister Lucy one from Watson Antrim & the other from M. P. Hargrove I forget whether I told you our camp is only 3 miles from the fortifications & pickets of the secessionists – I was not so well yesterday as usual I had a chill towards night yesterday but it was a light one I hope & think I will have no more I took a powder the Doctor gave me & I think it was that that caused me to be so chilly – our

This is part of the September 11, 1861, letter to Vashti.

Fort Ellsworth, Alexandria, Virginia; unknown photographer, formely attributed to Matthew Brady. Courtesy of the Metropolitan Museum of Art.

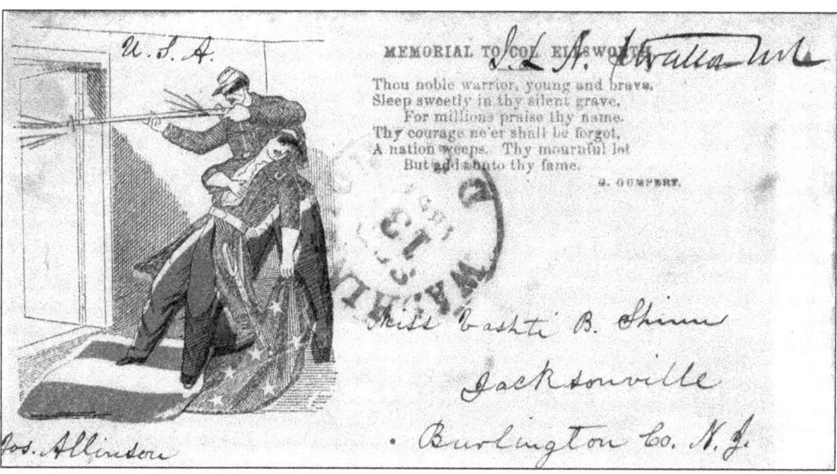

Commemorative stationary was often used by both soldiers and civilians. This envelope with a letter sent to Vashti memorialized the martyrdom of Col. Ellsworth. It has no stamp but is signed by Allison's post office friend Stratton.

graph office and secure the city of Alexandria, Virginia. On his way, Ellsworth spotted a Confederate flag defiantly waving from the rooftop of the Marshall House hotel.[132] The hotel's proprietor was an ardent secessionist, who raised the banner a month earlier vowing that the flag would come down over his dead body.

Incensed, Ellsworth climbed the three stories to the attic, cut the halyards and pulled the flag down. As he and his men descended the stairs, the proprietor stepped into the darkened stairway with a shotgun and shot Ellsworth, killing him instantly. One of Ellsworth's men retaliated by fatally shooting the innkeeper.

The union-wide response to Ellsworth's death marked a new direction in how Americans mourned and commemorated their dead.[133] As seen in Joseph's letter, visitors flocked to view the location of the incident. Stationary and other souvenir trinkets were issued.

❧

This next section appears to be some kind of diary, perhaps the "memorandum" he refers to in some of his letters. The entries are in installments (inst) and are pages of the same letter.

No. 4   Seminary near Alexandria, Virginia
September 25th 1861
Left Seminary & sick list today — Our 1st N.J. Regiment went out from our encampment on a march behind music & our head colonels it is said to be reviewed by Gen. McClellan — It is a beautiful clear day — I was on guard yesterday up in the cupola on top of the Seminary

Thursday 26th inst Thanksgiving Day
I left hospital yesterday there is preaching or church 3 times today & tonight — No news, only it is said our whole regiment & brigade to march tomorrow to take possession of some hay & something else that is now or will be in the hands of rebels — I must not have it very reliable

"there is preaching or church 3 times today & tonight" — Joseph Allison

On Sundays, command exempted the men from the daily drills and other military activities. Instead, they were offered the opportunity to attend religious services. While not mandatory, those who did not attend "have to stand two hours and a half and have the articles of war read to us which is not very pleasant."[134] Allison and most of the others understood the message.

Friday 27 inst

There was a gentle rain near or quite all last night — no rain till night — I was put on guard this morning 13 of us out of each camp. It has rained gently from early this morning till now (10 o'clock A.M.) This rain is likely to cause our Officers to abandon the design we had in view to make an advance movement upon the enemy — Yesterday was regarded here as elsewhere as a day of thanksgiving & prayer — There was a meeting in the chapel at the Seminary morning after noon & evening — The evening meeting was a Union as well as a religious meeting & quartermaster Sam Read of Mount Holly in the N.J. 1st regiment made the best Union speech that ever was made & there were a number — After hymns were sung the Star Spangled Banner was sung in conclusion & it was well done — There is an organ used in the chapel & the young man that plays on it is certainly master of both vocal & instrumental music — I must learn his name & get acquainted with him — The speech of Mr. Read I hear very highly spoken of by all & I hear strangers to Mr. R & his co. very much in his praise

Saturday 28th Inst

A great deal of rain fell yesterday I come off of guard today & they wanted to put me on for another day but they could not come it. It cleared off in the night (last) cold today a fine day, our whole regiment & other companies if not regiments went too went to day to visit the enemy & get some hay our government wanted to use — Our troops went to keep off the rebels while those with wagons hauled the hay — The secessionists run & left their johny cake silver dishes beds &c in their houses, in one house in particular corporal in my tent got a good lunch on caked eggs &c. Some of our camp D got some silver dishes of some kind & some one or more of our troops threw a clock of a mantlepiece & destroyed it the only excuse for him he may have thought it kept secession time — I did not go with our comp as I was on guard — There was some shots fired between our troops & the rebels & I am told one man in 1st N.J. & 2 or 3 in 4th N.J. Regiments were wounded.

Sunday 29th inst

A winter night last night almost — Our comp got into camp about an hour after dark last night — We have the promise of another beautiful clear day, the sun is warm & the wind is very cool Mailed a letter to T. B. Gaskill — It is whispered our Regiment is to go after more hay today but I doubt it — We have had roll call at 6 O'clock in the morning which has been a little after sun rise This morning we had roll call before day light by candle light which made us at the time think we are in for hay or something else if we go we are not going to start very early — I was not very well yesterday but feel extremely well today — Our comp & every other in this R. had an inspection of knapsacks this morning The cap. of each comp. makes the examination — There was also an inspection of our muskets & rifles this morning by our new Col. Commanding this Regiment Montgomery was promoted & we had to have a new man in his place — There is a report now that 2 men in N.J. 2nd R. of our troops were killed last evening by rebels

Tuesday October 1st 1861

Had no time to write yesterday — 50 men of our comp. D went yesterday to capture some wood from the rebels on the rail road from Alexandria to Richmond — We got 200 cords of wood good dry rack about 6 1/2 miles from here — We walked to the railroad then took the cars & rode all the way to where the wood is corded alongside of the rail road — We got wood in gun shot from Mr. Edsel who was born in Jersey & lived in Sussex Co. he is a Union man & gave us a great deal of information — Our troops were hauling hay too from both sides of the rail road where we got our wood — I got a coffee supper with Henry Clevenger last Sunday — It was very cold again last night very beautiful clear day yesterday & Sunday both — appearance of its being the same today — Our cap. Went to Washington last Sunday had not returned late night so our 1st Lieutenant took us yesterday & we had our 2nd ride in Virginia — There were two long trains went after wood yesterday we were in one train & loaded it full with wood & rode back in the other train — 1st train open cars for wood or the like latter freight cars but even they were loaded with some wood — We yesterday had a great deal of work to do to load cars in time not to be after night so we had not much time to listen to information from Mr. Edsel & others but I heard we were near Munson's Hill[135] — Sergeant Woodward says we were not near it — There are mile posts on the rail road & we got wood between 7 & 8 miles from Alexandria, Mr. Edsel says the rebels had an encampment there had some nigger[136] soldiers (I do not say these black soldiers were in camp by Edsel's I do not know about it) & the rebels were from North Carolina Mississippi & somewhere else he said — The rebels stole 30 or 40 cattle from Mr. E & they also took all his hay & I think something else — Rebels had Mr. E. prisoner 12 days — There was a guard on the rail road all the way we went & I do not know when I have seen as good looking soldiers — their countenances & eyes & actions beamed with intelligence bravery genius & thoughtfulness — our troops got a great many Baltimore papers (Clipper) as we started at the Seminary & our troops threw them out to the guard as we passed in the cars — We have a brass band now & we have the good Of it they are mounting guard now & they are playing Yankee Doodle so my ear is charmed while I use my pen — It is my favorite tune — mounting guard is getting out a new guard at 9 O'clock A.M. every day to serve till 9 O'clock next morning — new there are 13 taken out of each comp. in our regiment so that in all there are 130 men on regimental guard & they surround our whole encampment & regiment to guard it — Our troops drove rebels away from Munsons Hill & took possession of it last Saturday evening. I saw 2 boys living in Alexandria last Sunday one has one brother & the other 3 brothers in the rebel army — One told me his name is Howell & that his brother was in rebel army at Bull Run & was now at Munson's Hill — He need not have put it in the present tense because Sunday at that time *I mean before then* his brother had made good use of his legs to leave Munsons Hill — We had 4 horses to a wagon hauling hay yesterday for our government — This wood & hay we get belongs to secessionists — sent a letter to Mother. Barney Devlin's finger has got entirely well, he shot it with his pistol

Wednesday 2nd inst

Rain today but did not rain hard — Some of our troops in 3rd N.J. Regiment went out on scout yesterday not very far from Springfield

Bailey's Cross Roads Advanced Post of the United States Army Opposite Munson's Hill. *Harper's Weekly*, October 5, 1861, 629. www.sonofthesouth.net.

Station on rail road from Alexandria to Richmond — This road goes through Springfield which is 9 miles from Alexandria — said rail road from A. to Richmond is not finished all the way — Our troops went out near rebels who fired upon our troops & today 4 men are in the hospital of U. S. soldiers 3 of them shot in arm & one in leg — It is said a Doctor in 3rd regiment is missing these 4 wounded men are all out of 3rd N.J. Regiment — It is said several are missing in 3 R. it is not known certainly that any were killed on either side.

Thursday 3 rd inst

Beautiful clear day this morning — I see the troops in 2nd Regiment all moving towards Alexandria yesterday afternoon it is said they have removed their encampment somewhere else if they have moved it is somewhere the direction of Alexandria they have gone — Wm Cox of Pemberton N.J. is one of the 4 wounded men he is in Company C. Cap Rowand — Wm says he shot at the rebels with a black skin or black heart or both 3 times. Sent a letter to B. A. Antrim & one to Jason Cox — Sent a letter yesterday to Job. H. Gaskill D. F. Gibbs & squire Rogers — one letter to all The rail road we went on to get that wood being our end ride in Virginia from Alexandria to Richmond —

Friday 4th inst

Received a letter from Asa R. Foster — come off of guard at 10 A.M. 7 very unwell through the day — There are only 4 men in U. S. troops wounded last Tuesday Wm Cox of Pemberton in Cap Rowands comp & said they were told some of our troops that by a certain place the rebels had carried 3 dead men & 5 wounded ones — This engagement last Tuesday was near Springfield Station which is on rail road from Alexandria to Richmond though the road is not finished all the way — S. Station is 9 miles from Alexandria — The United States government use this rail road end next to Alexandria & rebels us it at other end next to Alexandria — clear day & very very warm

Saturday Oct. 5th 1861

Appearance of another beautiful clear day — 19 of our men going to Fort Taylor today to work I am now in — I took some medicine yesterday afternoon & I feel much better for it. All in our comp. D had the privilege of going to wash this forenoon in the creek a company of us went & I went the water was exceedingly warm for this time of year I halted on my return at Henry Clevengers where I was treated to a good coffee dinner — We had to clean all our brass today to be ready to be inspected tomorrow morning — Bought much the best song book I ever owned. Gave 15 cts for it 3 cts more than its price as the black man gave himself considerable trouble to get it It has between 75 & 100 songs in it the very best.

Sunday Oct 6th 1861

Met Wm Cox that was shot in the arm (from Pemberton N.J.) going from hospital to cap. Rowand's comp. on a visit it is the comp. Wm belongs in — I am glad to see him so well his Father grand Father & of his numerous relations every one have always been the strongest kind of Democrats having kind of inherited it I regret that one of the very same Democrat party should at all injure one who has acted so brave noble patriotic & wise as Wm Cox Jr. has done — His very many relations always acted with all their might against the old line Whig & Clay parties & even of late enjoyed all their skill & votes against Col. Fremont, Abe Lincoln, & the Republican party In Wm coming out like a man & a patriot for the "Union, the Constitution, the enforcement of the taws & one flag & one government" let the poet & God Almighty in his Providence apply these words in their application to young Wm who but a boy —

> Who feels the lashes of an adverse hour,
> Finds them but means to waft him into power, As
> health to bodies bitter draughts impart, So trials
> are but Physic to the heart.
> The tract of Providence like rivers wind,
> Here run before us, there retreat behind,
> And though immerged in earth from human eyes,
> Again break forth & move conspicuous rise[137]

The ball of the enemy struck the bone in the right arm of the elbow of Wm & lodged there; it had to be cut out by the Doctors & they knew this would be very painful & wanted him to take something to make him insensible, he was so much of a man, a true brave soldier & daring patriot, he said "no, cut it out, he would look at them, he would rather take nothin'. Of course the Doctors let him have his own way — As I said 3 or 4 wounded, were all three shot in the arm — One poor man had to have his arm amputated yesterday above the elbow — it hurt him very much & he manifested much pain, he followed the example of the bold Cox not to be made insensible — Col. Commanding our Regiment had an inspection of our everything this morning from the buttons on our coats to our tents & contents in them — clear day have a head ache — Returning from wash yesterday fired my new Minnie Rifle off for the first time — Corporal in my tent was a witness at the outside window to the amputation of that arm. I was not there. I knew It was said it would have to be done. Received a letter from Cousin Vashti Shinn — The rebels with their black skin or black hearts & consciences or all were not smart enough to immerge young Cox in earth from human eyes Cox is not dead but living & he can & will yet again rise conspicuous to the sorrow of the secessionists. Wm H. Shrope was team driver — he has been changed to a soldier in our ranks & got his Rifle yesterday. Mailed a letter to A. R. Foster esqr.

Monday 7th inst

It is cloudy this morning though the sun rose clear — Our cap. Gave orders all that wanted shoes to give their names & their size — No news — There was a little rain early last evening but very little Our comp. soon expects to have to go out on Picket —

Tuesday October 8th 1861

Rain early this morning yet 19 of us went to work in trenches at Fort Taylor. I went & we had to work in the rain — Mailed a letter to Mother & one to Col J. W. Wall Monday — Corporal made a mistake about that German[138] having his arm amputated I saw him yesterday & it is not done — Had some exceeding hard rain last night I think the hardest we soldiers have had any night yet — Had to take some medicine from Doctor for chills

Wednesday 9th inst

Very cold last night like winter same this morning — Had to take some more chill medicine from Doctor Gordon — Last Monday had to pleasure & honor of having a visit from Lieutenant Wm H. Campion M.D. he is in Vincentown comp. & asked me to come & see him at Alexandria — Cloudy North & South, East & West, but too cold to rain — received a letter from Jason F. Cox & mailed one for Joseph Carr Jr.

Thursday Oct. 10th 1861

Cloudy again this morning & cold last night — No rain yesterday — There was a drill yesterday forenoon & afternoon both — Wm H. Shrope went on guard yesterday for first — It was said our comp. is to go out on out Picket today to stay from comp. 3 days but Sergeant Woodward says Cap. Mutchler says we are not going — No news — Taking quinine for chills — Was requested by the Father of Wm Cox wounded in hospital shot in right arm to write to him (Jason F. Cox Pemberton) about certain matters & about his son. Did it today & mailed letter.

Friday Oct. 11th

Our comp. & comp. E went out on out Picket this morning I hear to Masons Hill to guard it against the rebels. The Captains Lieutenants & Officers of this Regiment had some kind of a meeting at Alexandria yesterday they did not return to camp in time or our comp. would have went out on Picket yesterday — Rain yesterday & somewhat cloudy today

"We were 3 months or more without any band of music we now have one of 17 pieces & we use it 4 or 5 time a day sometimes"
— Joseph Allison

"We have a brass band now & we have the good of it they are mounting guard now & they are playing Yankee Doodle so my ear is charmed while I use my pen — It is my favorite tune"
— Joseph Allison

"Our 1St NJ. Regiment went out from our encampment on a march behind music"
— Joseph Allison

"About 5 o'clock P.M. our bands commenced playing Yankee Doodle, Hail Columbia, Star Spangled Banner and a number of other patriotic airs, which served our men to such a pitch of enthusiasm that they cheered vociferously, and rallied on the enemy, causing them to retreat, and finally terminated the battle."

Camp Correspondence from near Williamsburg, VA, May 7, 1862 (*West-Jersey Pioneer*), 1.

During the Civil War, command required each regiment to organize a field band. Often, band members comprised boys as young as 12, playing fifes and drums. The bands raised spirits and promoted patriotism. In addition, they served to aid troop movements by conveying instructions and instilling order through their music. Their importance often exposed the musicians to snipers. Albert Woolston, the last surviving Civil War veteran, died in 1956 at the age of 109. Serving as a drummer boy, he lied about his age to enlist.[139]

∽

Headquarters 1st Regiment Seminary Camp Fairfax Virginia
Tuesday November 12 1861
Dear Cousin

I received your acceptable letter I think last Sunday it was not till some days after its date — I got a letter from sister Lucy at noon the 12th of last month & it was mailed at Pemberton the morning before — I think the cause of the difference was you directed your letter "Camp Seminary" & Lucy directed hers to the town of Alexandria — A letter will come to me sooner thus "I.A. Alexandria Virginia Company D 1st N.J. Regiment Care Captain Mutchler" Our camp is camp Seminary — With your printed envelopes you can put Alexandria below Camp separated from it so as not to read "Camp Alexandria" — Our Captain left our camp this morning to go to Philipsburg Warren Co. N.J. where his home is on a visit he will return about Friday — We were 3 months or more without any band of music we now have one of 17 pieces & we use it 4 or 5 time a day sometimes — We had a knapsack drill & review yesterday before Gen. Kearny in the large field by Fort Murphy — Next Thursday I learn we are to have a review or inspection by Gen. McClellan — We now have our dress parade in the morning till within a week we had it about sun set — We had our pay day last Friday — We were paid for 2 months up to 1st of this. Privates got $25 being $13 a month — Officers got more — You speak of one of your brothers having some notion of coming into our army[140] if he will come he can get into the company I am in as so many in our comp. has been promoted that there is any opening for him, my advice to him is to come by all means I think we can put this rebellion down without much blood shed I have made up my mind it must be put down any how with "Henry" give me "Liberty or give me Death."[141] We cannot die but once & in the order of an all wise Providence we cannot & will not die till our time comes — In our whole 1st N.J. Regiment we have only buried 3 or 4 soldiers think of it for 5 months 22nd of last month & 1000 too that does not look like our having to die only one of our soldiers in our 1st N.J. R. has been killed by the rebels. I was put on guard this morning. Our company was put on Picket at Clouds Mill we had a fine & easy time there — Our comp. returned to camp last Wednesday afternoon — Professor John L. Mountain Baloonist has two baloons at Clouds Mills & he has 3 men & 4 horses to assist him to get ready to take a aerial voyage or ride — Our camp here is upon the side of the hill upon which the Seminary building are erected the hill is 250 feet above the tide water of the Potomac we have a beautiful view from here of the broad Potomac river & we see the boats going up & down — We also have a good view of Washington & Alexandria both — We are 8 h miles from former & 2 1/2 miles from latter place — I expect it is Cousin Joe[142] that you say has some of the war spirit in him — We had some good music last evening & very very many cheers on account of good news we heard about the capture by our troops of 2 or 3 rebel Forts — Our band gave us hail Columbia Yankee doodle & some excellent tunes — I was at Henry Clevengers last Sunday — If your brother come here the sooner he comes the better — We will get $13 a month & plenty of clothes without having to pay for them — I got a new pair of Pants last Sunday that is 3 pair I have got from Uncle Sam We expect to get a very nice dress coat tomorrow they are like those the 3rd & 4th N.J. got so we have seen the kind — The teams went to day to Washington after these coats — We each got a new cap about a week or 10 days ago — You often speak of the traitors[143] in Jersey I am sure as you say an example should be made of some of the worst & if a few were hung it would have a good influence to put down rebellion — do not know what it is some men can think of or why it is they will go against their own interest the Union the Constitution the enforcement of the laws & one flag & one government. Some care not what they do & it has been of late years as Henry said of a part of the Democrat party "They woo but to ruin & win but to destroy it." For my part I desire to have a more permanent & steady rule for my conduct the dictates of my own breast. Those who have forgone that pleasing adviser & given their minds up to be the slave of every popular impulse I sincerely pity. I pity those Still more if their vanity leads them to mistake the shouts of a mob for the trumpet of fame. Experience might inform them that many who have been saluted with the hurrah of a crowd one day have received their execrations The next & many who by the popularity of the times have been held up as pure & spotless Patriots have nevertheless appeared upon the historians page when truth has triumphed over delusion the assassins of Liberty. We see now what the delusion of some has brought them & us & our country to — We intend that truth Freedom Justice Humanity & Independence shall triumph.

12th Inst

I have only time to add a few lines —We had professor L. Mountain with his baloon named Saratoga in our camp to day he come down near us & then went all through our camp the height of a mans shoulders some men having to hold him down by means of ropes he would have went up if he had not been held down another gentleman was in the basket with Mr. L. Mountain — We are today making ready for the Review tomorrow — Give my love to all our relations — I have not received any letter from Louisa Ward or Lizzie Foster tell them to hurry them up —

Affectionately your cousin Joseph Allison
Miss Vashti Shinn,
Jacksonville N.J.

"Professor John L. Mountain Baloonist has two baloons at Clouds Mills & he has 3 men & 4 horses to assist him to get ready to take a aerial voyage or ride —"
—Joseph Allison

"We had professor L. Mountain with his baloon named Saratoga in our camp to day he come down near us & then went all through our camp the height of a mans shoulders some men having to hold him down by means of ropes he would have went up if he had not been held down another gentleman was in the basket with Mr. L. Mountain —"
—Joseph Allison

Balloons in the American Civil War

Both the Union and Confederate armies used balloons for reconnaissance during the American Civil War, marking the first time that balloons were used in the United States for reconnaissance. Thaddeus Lowe and John LaMountain both carried out reconnaissance activities for the Union army during the war. The editor of the *Cincinnati Daily Commercial* wrote to U.S. Treasury Secretary Salmon P. Chase and suggested that the United States establish a balloon corps under Lowe's command. This corps would provide aerial reconnaissance for the Union armies. Aeronaut John LaMountain was also attempting to provide balloon services for the Union. He wrote to Secretary Cameron in 1861, but, because he had

John LaMountain attempted to provide balloons for reconnaisance to the Union troops during the U.S. Civil War.

no influential backers, LaMountain did not receive a reply. However, the commander of the Union Forces at Fort Monroe, Major General Benjamin F. Butler, contacted him and asked for a demonstration.

LaMountain's reconnaisance balloon.

The New York Times reported that (from Fort Monroe) LaMountain could view the Confederate encampments beyond Newmarket Bridge, Virginia, and also at the James River north of Newport News, the presence of the balloons forced the Confederates to conceal their forces. To avoid detection, they blacked out their camps after dark and also created dummy encampments and gun emplacements, all of which took valuable time and personnel; however, the balloon corps did not last until the end of the war."[144]

LaMountain and his balloon performed important reconnaissance work in advance of the battle of Fredericksburg. Joseph's brother John would have observed it as well. The John L. Roebling Company of Trenton, New Jersey, supplied the wire cable for the balloons. In addition to the balloons, the Civil War created a big demand for cables to build bridges and ships.[145]

"We had a grand review last Thursday it gave me the head ache & I have not got clear of it yet. . . . At our Review last Thursday we had somewhere between fifty & seventy five thousand — Very many say latter but (5000 1 think) is about the mark — We were reviewed by Gen. McClellan, President Lincoln, Secretary Cameron, & very many other distinguished characters."
— *Joseph Allison*

"After the Union defeat on 21 July 1861 at the first Battle of Manassas, Lincoln appointed Maj. Gen. George B. McClellan as commander of the demoralized army. A superb organizer, McClellan rebuilt the army and on 20 November 1861 staged a formal military review (at Bailey's Crossroads, Virginia). . . . Lincoln and his entire cabinet attended. Occupying nearly 200 acres, some 50,000 troops . . . took part in the review, at that time the largest ever held in America." *Historical marker formerly marking Bailey's Crossroads.*

Camp Seminary Virginia

November 22 1861

Dear Mother

We had a grand review last Thursday it gave me the head ache & I have not got clear of it yet; I tried to get Doctor Gordon to excuse me for duty today, though he gave me (I mean Dr.) 4 pills & said after it I must take castor oil I took the pills thee sent instead yet Gordon said he had it not in his power I went to Major Hatfield & Mr. H could not do it & said Doctor was only one that could do it — At our Review last Thursday we had somewhere between fifty & seventy five thousand — Very many say latter but (5000 I think) is about the mark — We were reviewed by Gen. McClellan, President Lincoln, Secretary Cameron, Gen Kearny & very many other distinguished characters — I intend to have a furlough some time next month & if I cannot get one of our Superior Officers to get me one I will write to President Lincoln or Gen. McClellan & I am sure I must succeed in the end — I have been in this company 6 months today & as a single man I can get $24 state pay but I could not get it till I returned from the wars — If I get killed as a matter of course I would never get it I see a soldier with a wife or having a widow Mother[146] can draw $5 a month so that will make a difference of $12 up to today & another advantage is we can have the use of the money & use it in such a way as to make it pay a good interest — Thee[147] can draw $35 I think, I have not said any thing to our captain about it but I am going to when or before I speak about my furlough — If I get no furlough or mean

This savings coupon was in one of the letters.

prospect of getting it under 3 weeks or more (I think I will have to sign a certificate, I will sent it on to thee by mail & thee can draw it — I desire that scamp at Trenton[148] not to have the use of one cent of it to put it out on interest or use it in any other way I am willing — Thee knows we have one time been true friends to Cooper & we both know how he had treated us afterwards in regard to my final settlement with the execution[149] I paid some sheriff costs & cousin Samuel said they executors would pay all costs this matter was only a dollar & a few cents for each executor cousins Sam & Wm. Each paid me one dollar & twenty five cts & it was a few cents more than their share John paid me in board & when I was at Camp Olden[150] I called on Cooper to pay his share though I had not a cent of money at the time Cooper told me he would not pay it. I intend he shall pay it — We got good new tents yesterday & Gen. Kearney yesterday excused us from all duty XXX dress parade — We hear the roar of cannon & one of our officers in command says there a battle going on — get Lucy to tell me the number of my last memorandum be sure & do I do not know how to number the one I have now — Corporal in my tent has gone to Alexandria to purchase a stove exclusively for the use of No. 3 tent where I am now there was a change made to size us off & put those in a tent all of a height Armstrong is well & is now in tent No. 2 he was in No. 3 — We only got 2 tents for the Privates we had 6 before. New tents we got yesterday were made in Newark N.J. & were patented last June. Love to all — Affectionately they son

Addressed to Beulah Allison

— Joseph Allison

"as a single man I can get $24 state pay but I could not get it till I returned from the wars — If I get killed as a matter of course I would never get it I see a soldier with a wife or having a widow Mother can draw $5 a month so that will make a difference of $12 up to today & another advantage is we can have the use of the money & use it"
— Joseph Allison

Privates in the Union Army earned $13 a month, $2 more than their Confederate counterparts.[151] Beulah Allinson applied for, and received, Joseph's pay.

					(3-H-3)
NAME OF SOLDIER:	*Allison, Joseph*				
NAME OF DEPENDENT:	*Widow,*				
	Minor,				
	Mother	*Allison, Beulah*			
SERVICE:		*D, 1 N.J. Inf.*			
DATE OF FILING.	**CLASS.**	**APPLICATION NO.**	**CERTIFICATE NO.**	**STATE FROM WHICH FILED.**	
	Invalid,				
	Widow,				
	Minor,				
1862. Mar. 4	*Mother*	*6.744*	*467*		
ATTORNEY:					
REMARKS:					

Pension Record for Beulah Allison/Allinson, U.S., Civil War Pension Index: General Index to Pension Files, 1861–1934. National Archives and Records Administration.

❧

Pemberton N.J. 12 Mo. 22 1861

I did my duty today to get excused this afternoon but had finally to excuse myself with my bad head ache & I want to see them try to force me on duty. Our Col is determined we shall look well he is very proud himself — best pants we just got another pair similar each & every one of us

❧

Dec. 4, 1861

Dear Sister

I received thy letter & bundle through Ben Nutt Abram says it was not the kind of shirt he wanted I thought thee knew I complained to thee that the government give us plenty of these under shirts but what we want is an over shirt to go over the under shirts we get from the government — Those socks are very acceptable we have been waiting & waiting for the government to get us some socks but they have not come if Mother has good chance soon to send me another pair do it if not I can do with out them Abram & I do not want any more of these shirts I will try to exchange this for an over shirt — I will have a furlough between this & the 10th of next month & I can make out with the socks & shirts I have got — I wish Mother would give Joe Hargrove $2 for me I loaned that amount of him before I left Jersey I wish John would pay Joe Carr $7.50 Cts for the Mount Holly Mirror & take Receipt in full for one year from the 22 of last April or 1st of last May — Thee does not tell me one thing or the other whether John will collect the money for N.Y. Weekly Tribune I think I gave the names of those that will be sure to take it I got a letter from squire Rogers today & he way they are looking for Tribune & want it as it has stopped the 25th of last month Mr. Rogers gave me these names as wanting Weekly Tribune

1. Japhet Alston
2. Joe Butterworth
3. Theodore B. Gaskill
4. Same Gaskill
5. S. Morris
6. Wm English
7. D. Lippincott
8. Mary Lippincott
9. Frank Jones
10. Jacob Seeds

These Rogers says have given in their names & they are all good for the money I wish John would collect it I would be willing to divide the profit with him to these John Croshaw Joseph Croshaw Abram Alston A. H. Fort Joe Wells J. C. H. Combs Frank Egbert David Vandever Shreve Shinn Samuel Smith Mrs. J. S. Gaskill (widow of Job) Stacy Webb Charles C. Shinn Benjamin Burr Benjamin Lamb & Joseph Cox & the store keeper at Hanover Furnace all may be added as certain to take Tribune let no one but the widow of Job S. Gaskill have it for one dollar charge on dollar & a quarter & that is only 5 cts above its retail price but if John or Rogers send any money to Greeley they must only send one dollar for one copy — We in my tent have bought a good stove & put it in our tent & it keeps us warm very many wish they had one like it they have fooled their money away some have, so they have not the money to buy a stove we do not get potatoes now so it is quite a privation to us — I fare well but it is at my own expense in part — Give my love to all

    Affectionately   thy Brother Joseph

    Envelope addressed to Lucy A. Allison

Dec. 24, 1861

Dear Cousin,

I received your letter of the 15th of this month I was very glad to hear from you we have had some changes & news but the papers are best for us to look for the news — About the 13th of last month (I think it was) a man was shot by the name of Wm. H. Johnson[152] about three quarters of a mile from our camp here — Johnson had deserted us & went over to the enemy & was caught in the act & it cost him his life — Our whole Gen Franklin's division had to go out & see him shot we each had a chance to see him before he was shot — 12 men that were to shoot him marched round in front then a covered wagon with the coffin of Johnson then a wagon with Johnson in & a priest alongside of him this wagon had no top to it — I was about 12 yards from Johnson when the wagon was driven that distance from our Co. D. After J. was shot all companies marched by Johnson to see him after he was dead — It was done to be a warning to others not to follow his example — I had several friends to come & see me out of the Cavalry Co. got up at Columbus N.J. last Sunday also 2 or 3 out of 4th Regiment — I see Jonathan Fox[153] is deceased he lived in Pemberton — Cooper & John bought that farm of Fox— I got a letter from Louisa Ward not long ago — We are fixing our tents for winter by setting split logs end ways in the ground about 5 feet I length then mud is used as morter to put between the logs to keep the cold out. The bottom of our tents are then put on the top of the logs by means of wooden pins to hold the tents — The 4th N.J. R. have their tents fixed so & we follow their example — I have not been very well for some days I have a cold & head ache I took some pills this morning — I am going to try this week to get a furlough to make a flying visit of a week in Burlington Co. My cap-

Joseph Arey Shinn (1843–1918), Vashti's brother, was 18 years old when he received this letter. Two years later, he enlisted in the 3rd Regiment, New Jersey Volunteers. Photograph by Dalrymple, Photo-Artist, Hightstown, N.J. is in the Robbins family archive.

tain is willing for me to go & think it is likely I will get a furlough though very many are refused — It has been some time since your sister wrote to me think t wrote to her last. Is there any body in your

neighbourhood that would like to unite with us to put down this wicked rebellion? If so an opportunity now offers they can come into N. J. 1st in either co. they please — There is an opening in Co. D. for one or more — When I return from my furlough here I will bring all that want to come free of charge — Give my love to all my cousins & friends — Write as often as you can — My address is you can tell my friends is "1. A. Camp Seminary Alexandria Virginia Co. D. 1st N.J. Regiment, Care Captain Mutchler." Love to you all

Affectionately your Cousin

Joseph Allison

12 Mo. 24, 1861

Envelope addressed to Joseph A. Shinn

Jacksonville, N.J.

"We are fixing our tents for winter by setting split logs end ways in the ground about 5 feet I length then mud is used as morter to put between the logs to keep the cold out. The bottom of our tents are then put on the top of the logs by means of wooden pins to hold the tents."
— Joseph Allison

Camps featured many tents housing five or six men. This is a Union encampment at Cumberland Landing, Virginia. Courtesy of the Library of Congress.

The Union army also used log cabins in winter months. Chimneys would be built for a fire to keep warm. The picture shows an officers' winter quarters at the Army of the Potomac headquarters. Courtesy of the Library of Congress.

Camp Seminary Virginia U.S. Army Feb. 6 1862
Dear Sister

Washington Watts Wm Cox & I went to Alexandria yesterday & when I returned which was about 4 P.M. I got thy letter I thought it was about time as thy pen only has to answer for Mother John & thee all three it appears it only takes 6 months or more by thy letter for me to be informed of Mother getting my key I have not been informed before of it yet thee thinks I have forgotten it I have no such a short memory about such matters because I have for more than 6 months kept my eyes & ears open to learn where the key is & has been for a long time — My mother has the letters to me & thee can see whether any one tells if Mother had got my key I am very much mistaken if any one of thy letters give said news — Mother had all my letters & I send them all to her — Our last pay day was the 10th of last month & very soon after I bought a good watch for $13 it works very well by Ross team driver in our company offered me $16 cash for my watch I took it & bought another in Alexandria yesterday of W. W. Adam a native Virginian & an honest & reliable man. Mr. A. had been in the jewelery business ever since he was 16 years old I should guess he is now about 45 — I have bought a better watch this time & it costs $15 — I bought 3 pounds of sausage at Alexandria yesterday but they gave me at least 4 pounds I only had to pay 36 cts however & our mean sutler[154] 20 cts a pound I intend to buy no more of the sutler though our soldiers are such fools as to do & the sutler should not have so much encouragement — just sent a letter to Asa R. Foster— I think I have told Mother we have a stove in my tent which keeps it very comfortable for us soldiers there are 17 men in my tent — Abram is well — Wm Cox has sent a furlough to be signed for him to come home so you may soon see him as they certainly will not deny him a couple for life as the Doctors say he will never again have the full use of his arm on account of its being stiff in the elbow — Wm told me yesterday he can write now with his lame arm hope he will get entirely over it but fear he will not a piece of bone come out last Monday & Cox carries it in his pocket to show to his friends — Capt Mutchler went to Alexandria yester-day — We have not had 3 clear days all day for the last 3 weeks it was clear yesterday all day but we have rain again today I never saw such weather in my life — Tell Mother to tell Rachel Powell Hannah Clevenger says she is so lame in her foot she does not know what to do, there is a hole in it & it gives her no comfort— I & others fare very well but we have to pay for our living as I cannot live on the government fare we often do not get what the government pays for us to get we are allowed bean & rice we do not get these at all but they should give us potatoes & pork steak or sausage or something in lieu of rice & beans but we get nothing of the kind some come here to make money & they will make it they care not how they do not want the rebellion put down as then their big pay will stop — I am glad Hargrove & Carr have received the $3 1/2 due them — The Mirror told me of the death of Thomas Goodher & I think Abram told me too some body else did — 2 men out of my tent had gone to Alexandria today so many can go out of a company at a time but no more — it is George Emmons & W. T. Bennet — they wanted to go very much or they would not have went in the rain as they have— I have not taken time to eat dinner with the rest & I must hasten to a close. I am confident we will not in N. J, 1st leave here this winter to make an advance movement upon the enemy — Gen. McClellan was in Alexandria yesterday I see him so often I know him very well he has a very piercing penetrating eye & looks like a business man — We have a board floor in our tent one foot from the ground we can keep things under our board floor—A cat come into our tent took board in the lower story under our floor we thought such company would keep out the rats & mice said cat now has a family of 3 kittens I did think I would fry some sausage but we have some boiled fresh been & I will eat a piece I have got on my plate — I only eat some toast bread & butter & coffee for breakfast — I am very well satisfied & pleased with my new watch it is a companion very acceptable to me — It is expected our company will go out on out picket next Monday — I got 10 cts worth of puddings put up in the style of sausage — Love to all I expect Abram sends his love — The 22nd of this month I was about 100 yards & no more outside of our lines the patrol put me in the Provost guard[155] house on account of it our Capt was away but John Schoonover Capt's Clerk & Sam Read had me taken out in less than 48 hours afterwards — I had better quarters in the guard house the 2 nights I was there than we can have in our tents — Provost guard house is a good house belonging to the Seminary — it is lathed & plastered I was put in best room with a good fire — I got a letter from Peter Ellis he say he sold his port for $4. Our sutler in the way of sausage must have $40 a hundred for his port— I send a little present to Mother & thee I give 15 cts for all 4 of these men & will call it a military valentine — Write as often as thee can do it I intend now to fry some pudding we can do it on our stove — my meat has got cold

Affectionately thy brother Joseph Allison

Addressed to Lucy Allison

&

U.S. Army Camp Seminary Va

Feb 27th 1862

Dear Cousin

I received a letter yesterday from Jacob Laumasters & 3 others from cousin A. R. Foster one of which is from your brother Joe I am glad to hear you are all well — I regret to hear the death of Capt. Joe F. Rowand he was in Co. C & not in Co. H as the Phila. Inquirer said Mr. Laumaster says the funeral of my particular friend Rowand took place last Tuesday very many will drop a tear over his being taken from us in the morning of his days but we must seek consolation in the words & spirit of the past

> O wake not with mourning,
> the rest of the dead
> For the blessed in heaven
> No tears should be shed
> But weep for the living
> Who linger to bear
> The burden of sorrow
> Of anguish and care.
> In those who are sleeping,
> In peace & in love,
> Whose hopes were all treasured
> In heaven above,
> No more need our watchings,
> Our tear or our prayers,
> They have left far behind them
> Earth's troubles & cares
> Then rejoice that they've passed
> In the prime of their years
> From this world of anguish
> Of sorrow and tears
> Their glorified spirits unfettered & free
> In these heavenly mansions
> Forever shall be,
> Tis weakness to mourn
> Or wish them back here
> We would not recall them,
> From yonder bright sphere,
> Their bright weariless wings

Through heaven they sweep
And a fond watch of love
Above us they keep
O wake not with mourning
The rest of the dead
For the blessed in heaven
No tears should be shed.[156]

Joshua Norcross on the Newbold farm next farm to Ben Gauntt near Jobstown had two sons in the 4th NJ. Regiment Co. I in the Mount Holly Co. XX Guards — One of Joshua's sons died in the hospital here & was sent home only a few days ago — Co. C & our Co. went out on a scout last Saturday we got up at 2 in the morning & left here precisely at 3 o'clock as soon as we got breakfast We went to Bentons' Tavern which is 8 miles from Fairfax Court House. We did not find any of the rebels to shoot so we returned to camp towards night the same day — Two weeks ago our co. was on out picket at Bentons Tavern we were there a week one co. goes out at a time now & stays a week — It is said we have marching orders to that there is a probability of our leaving the Seminary here — I think there is no certainty about it as there has been so much said about marching from here & we have not marched yet — We have rain again last night & the mud is very bad to walk through — We have not had 3 entirely clear days & nights for two months — I went out this morning to get a warm breakfast which is richly worth the 25 cts I have to pay for it — In the army we only get bread meat & coffee & some times bread, meat & soup — The meat is always boiled as the cooks in our co. are too lazy to fry or broil any of our fresh meat — Last Tuesday I got dinner with Anty as she is called bill of fare was coffee with milk & sugar in it, fried fresh beef, roast fresh pork, a stew with fresh meat, potatoes & onions, fresh port cold, stewed oysters, stewed apple, mashed potatoes, boiled rice & cold wheatcakes in lieu of break — I was hungry & got the worth of my money. I found it very discouraging about getting a furlough so I did not even get one written to send in to be signed — My friend Laumaster said he heard I was coming home on a furlough so they had commenced to cook something to eat so I could have some nourishment when

I got to Jersey — We are mustered into service every two months & the time commenced from the time we are paid to —

The 10th of last January we were paid to the 1st of January that being due us from Nov. 1st 1860 to Jan 1st 1862. Some of our soldiers never shave so they have a long beard we have a brigade barber in the Seminary & get shaved there once a week I only have to pay 5 cts if I let barber comb my hair & colougn it he would charge 10 Cts — We have to clean our military implements today as tomorrow we are to be mustered into service as two month pay is due us & it is just 2 months since we were mustered in — Give my love to your Father Mother Cousin Joe your Uncle Asa & the rest of our relations I must hasten to a close — We have some sunshine today which is something new — I do not know whether you know we are in Gen. Franklins Division & the brigade of Gen. Kearny — Kearny only had one arm, he was in the war with Mexico was shot in the left arm & thus lost it. I have not taken time to get a better pen I must do it before I write another letter — My address is "U.S. Army Camp Seminary Virginia 1st N. J. Regiment Co. D. Care Capt Mutchler" Love to all my friends. I am very

    Affectionately
    Your Cousin
    Joseph Allison

    Addressed to Miss Vashti B. Shinn
    Jacksonville
    Burlington Co. N.J.

The First Moves Out

In March of 1862, the 1st Volunteer Infantry was attached to the 1st Brigade, 1st Division, 1st Army Corps, Army of the Potomac. The Union forces began advancing toward Manassas, Virginia, to join the Peninsula Campaign, the first large-scale operation in the Eastern theater of operations. Allison now found life very different from his Alexandria encampment stay.

Cedar Creek & Cattlet Station Virginia
Thursday April 10 1862
Dear Cousin

I have taken a seat on the ground with my overcoat woolen & Indian rubber blankets for my chair or seat I have taken down my small tent to dry it & let the sun dry the ground under my tent as we have to lay on it at night thus I have no covering over me but the blue canopy of Day but the sun is warm very different from what it has been with in a few days — Our 4 first N.J. Regiments are all here & we all left the Seminary last Saturday & came on the cars over the Alexandria & Orange rail road to Bristoe Station Prince William County 32 miles from Alexandria & the same distance from our camp at the Seminary we expect we have bid adieu for a long or forever to the Seminary so we brought our small tents & all our military rig we camped Saturday night at said station & Sunday afternoon we left & come here where I date my letter — We come here on foot as the rebels had destroyed the bridges cars & rail road between here & for some distance towards Alexandria & between here & Bristoe Station the road & bridges were not yet repaired they are now, so that we see the iron horse here & I am sure they have been smart to so soon repair the road & bridges between here & Bristoe Station (distance 7 miles) since we left the latter place on Sunday last — The bridge over the creek was also destroyed by the rebels & our Government commence building a new one yesterday it will cost a great deal of money as the creek is very wide & deep and the rail road is upon a very high embankment so that a high or deep & long bridge is very expensive & difficult to build — It is 7 miles from Bristoe Station to this creek & it is 2 miles to Warrington Junction from this bridge It is only a few hundred yards from Cattlet Station where the cars stop to this bridge there is a beautiful stone wall on each bank of this creek at this bridge the stone looks like unfinished marble I was on guard at the bridge last Tuesday I measured one stone & it is ten feet long some are longer but they were out of my reach so I could not measure them — I have been writing a letter to my sister Lucy & another to Joe Carr. Editor of Mirror for him to change address of my paper from Alexandria to Washington — I think you write to me last & it may be you have not written for some weeks because I had not answered your last letter I have many duties cousin & have much upon my mind so that you & others must excuse what you may call negligence — I want with others to have the pleasure of marching into Richmond & I presume this is our object & aim in our advance movement upon the enemy — I have my tent to put up & my daily memorandum to notice with other things to do that I must hasten to a close — I forgot whether I have yet told you in my letter about the cold winter weather we have had ever since last Monday noon — Last Monday at T.P.N. we were going to march on but a very violent cold snow storm come in & there was no stop to it till after 10 o'clock last night — I cleared off some time in the night — We had rain snow hail & very cold winter wind for 2 ½ days & 3 nights it rained snowed or hailed (which was after mixed) more than ½ of all that time night & day — We had a fire in this open field in fact very many of them yet it was so cold we burned our shoes socks & clothes to warm on one side wild the other side was cold enough to freeze — I thus burned my socks & pants & fear my shoes but we were so cold it drove us to very abrupt & extreme measures. With this warm sun hope an Indulgent Providence will smile upon us in our efforts to put down this unholy & unjust rebellion & make sure to one more have the Union the Constitution enforcement of the laws & one flag & one government again established even in the secession states — Our & my compromise may be nothing more but is mud sills will not condescend to accept of the least thing less we know our rights & knowing we dare & swear we will maintain them. This fire, famine, sword, desolate fields, desolate towns & thousands scattered abroad the south & rebels have brought upon themselves to themselves their country their wives their children & their God they will have to answer. The time for words with them (& argument too) has passed the present is now the time for action. We are not unhappy here, cousin, if we are on the damp ground, at the desolate & destroyed bridge the rebels will feel & see & know who most shall feels the lashed of an adverse hour — The rebels are not smart enough to immerge us even with their cannon & masked batteries, in earth from human eyes, we will show them who shall rise most conspicuous — Our pay day for January & February was last Friday — day before we left the last day of this month $26 will be due each Private—We had a good & glorious time at the beautiful Seminary but we expect our advance movement upon the enemy will be much & more glorious — Give my love to your Father Mother Cousin Joe all the rest of you & the whole of A. R. Fosters family. Give my love all my friends

I am your Affectionate Cousin Joseph Allison

Addressed to Miss Vashti B. Shinn Jacksonville
Burlington Coe N.J.
Change my address to this "J.A. Co. D. 1st N.J. Regt. Col Torbert Commanding Care Capt. Mutchler Washington D.C."

"Our 4 first N.J. Regiments are all here & we all left the Seminary lost Saturday & came on the cars over the Alexandria & Orange rail road to Bristoe Station Prince William County" — Joseph Allison

"The bridge over the creek was also destroyed by the rebels & our Government commence building a new one yesterday it will cost a great deal of money as the creek is very wide & deep and the rail road is upon a very high embankment so that a high or deep & long bridge is very expensive & difficult to build — It is 7 miles from Bristoe Station to this creek & it is 2 miles to Warrington Junction from this bridge It is only a few hundred yards from Cattlet Station where the cars stop to this bridge there is a beautiful stone wall on each bank of this creek at this bridge the stone looks like unfinished marble I was on guard at the bridge last Tuesday measured one stone & it is ten feet long some are longer but they were out of my reach so I could not measure them" — Joseph Allison

❧

York County Va. Tuesday   April 29, 1862
Dear Cousin

I recd you letter last evening & another from cousin Abram at Burlington I also got my mirror If last week & week before last as we have not had a mail for near 2 weeks & only one since we left Alexandria — Our mail goes & comes from here to Fortress Monroe — Your

Orange andAlexandria Railroad. Photo, Andrew Joseph Russell (formerly attributed to Mathew Brady), The Metropolitan Museum of Art.

"The O&A allowed farmers to ship their products, produce and goods much more cheaply than before. As a result of being the northern terminus of the railroad, Alexandria became a thriving seaport and manufacturing center."

"During the Civil War, the O&A was arguably the most fought over railroad in Virginia. The railroad provided the most direct all rail route from Washington to Richmond and the Orange and Alexandria would serve as a main highway for the troops to march on and be supplied. The South vigorously defended the railroad against this invading force with the result that several major campaigns (First and Second Manassas, Bristoe) and dozens of battles and smaller engagements took place on or near the tracks of the O & A. The rail junction at Gordonsville would play a particularly prominent role in Robert E. Lee's operations throughout the war. So important was this junction that it can probably be stated that nearly every soldier in the Army of Northern Virginia passed through it at one time or another."[157]

letter come 9 days after its Late it will be a good way to do as you do in your letter & that is say how soon we get a letter after its date — We cannot write often now there are not 10 men in my co. now that has any ink Captains & lieutenants not excepted — Another thing our mail carrier does not keep postage stamps just now & we expect soon to go to Yorktown — Cousin in your mournful story about this lady & her brother why is it you do not give me their names? It is in that order of Providence that they should be separated as you say but in our condolence for them let us not approach to melancholy on their account as Whittier[158] of Massachusetts says "Before the joy of Peace, must come the pains of purifying. Endure & wait & labor" We landed here 8 miles from Yorktown in York Co. near Perquasin[159] river last Tuesday the 22nd so it is a week today — My captain had just given Sergeant Wood orders to write a descriptive list of our company that is to put on paper of each man color of his hair eyes complexion XXXX so when any one is lost to have some guide to find him — We are asked if we are married or single but are not questioned as to whether we are engaged or in love — It is said it is now contrary to law for our sutlers to follow the army so our sutler Brown & Coe has not been with us since the 5th of this mo. when our reg. left Camp Seminary & camped at Bristoe Station in Prince William County Va. On the Alexandria and Orange railroad 31 miles from our camp & Alexandria we went in the cars there & next day 6th April went 8 miles on foot to Catletts station & Cedar Creek both places of which are in Fauquier County — These two last places are only a few hundred yards apart we were at these last places from Sunday 6th till late in afternoon of Friday April 11th then we come on foot to Bristoe Station & 12th we walked as far as Fairfax Court House then Sunday 13th of this month we walked from Fairfax to Alexandria & camped at Fort Ellsworth which is on the out skirts of Alexandria we remained there till Thursday week last when we left to advance upon Yorktown — Our 1st N. J. Reg. sutler having left us a Dutch sutler turned up on the boat a day or two before we landed here this Dutch sutler sells eggs at 50 cts. a dozen which is between 4 and 5 cts apiece; he sells cheese at 50 cts a pound & pigs feet at 15 cts each —We had rice for dinner today have not had any for a long time now I intend to ask our Col if he knows how this Dutch imposter is doing business & if can drive or turn him out of our Regt or bring down his prices I intend to do it. I will send you a card where I bought a watch in Alexandria & another where I bought something else they are of no use to me now as we have left Alexandria for a long time & perhaps forever — Charles Sitgreaves at Philipsburg Warren Co. N.J. the father of our first Lieutenant is talked of for Gov of N.J. on Democrat side[160] — But no one of that style can succeed to come in our Lieutenants Father is a good man & had done more for our Co. than any other man but his defeat either of the nomination or of the nomination and election both will be a verdict of the people not against the man but against the party & company he is in & had been — If a Democrat must & shall have it then I would say let Major Charles Sitgreaves have it am sure his son our 1st Lieutenant is a true gentleman & the Major at Trenton brought our company something to eat a wagon load at a time Major S. has been here to see us even in Virginia — he had showed us more favor than any body else yes in the Co. or out of it — I see by Mirror eggs in Mount Holly are only a cent apiece — We have salt water in the creeks here we are not very far from the York river and Chesapeak bay — We can get oysters & clams here in the creeks — The river & all this navagable water is alive with steam boats & our soldiers they extend far & wide much farther than our eye can reach them — There are so many boats we could not number them even to go into that business exclusively — The boat Hero brought our whole Regt. Here — The Elmira City boat brought another part of our brigade — The John A. Warner from Burlington N. J. that took stock from there to Phila. has been chartered by Uncle Sam it come with us here & brought the 3rd N.J. Regt. The Daniel Webster also come in our company I presume with a part of our brigade — We often hear the cannons roar here it is truly like war but we have come here prepared for the worst we only want to have our camps trimmed & our lights burning we know a day or an hour of virtuous liberty is worth a whole eternity in bondage I would rather therefore die in a good just right cause where we are than live in Confederate states with rebels in their unjust unholy combinations against Liberty Freedom Union & Independence It was this our forefathers fought & bled & some (very many) died for to bequeath & hand down to their children & their childrens children equal rights equal laws Union & Independence an inheritance uncorruptable undefiled & that fadeth not away — The fences houses niggers[161] every thing of the rebels is lost to them by their own bad management but we Union men living & dead both now & forever intend to have an Independence to celebrate upon the 4th of July & we do not intend that it shall be temporary — We had a clear day in part but we are now having some rain — I must conclude — I am glad to hear from you often your letters from our homes are a great consolation & delight & pleasure to us — Give my love to all my relations & friends especially to your father Mother Uncle Asa & family & my cousins in Burlington — We in the army do not want any of you to mourn over our situation in the Providence of God we are here one in spirit bone of one bone flesh of one flesh & it will & shall be so when the mortal shall put on immortality & when Death shall be swallowed up of life — Therefore we will not live with rebels & enemies to the Constitution they shall succumb & bow & yield to us or else we will fall & die at their feet beneath their power — We know we are not hurt yet & they have made such good use of their legs to run away we are not frightened — I do not think then your Good Bye is Farewell forever — Write soon

 Affectionately Your Cousin Joseph Allison
 Addressed to Miss Vashti B. Shinn Near Burlington N.J.

"The river & all this navagable water is alive with steam boats & our soldiers they extend far & wide much farther than our eye can reach them — There are so many boats we could not number them even to go into that business exclusively — The boat Hero brought our whole Regt. Here — The Elmin City boat brought another part of our brigade —The John A. Warner from Burlington N. J. that took stock from there to Phila. has been chartered by Uncle Sam it come with us here & brought the 3rd N.J. Regt." — Joseph Allison

On land near York river Virginia Friday — May 8 1862
Dear Mother

 We have entered the field of battle now in earnest we saw yesterday & day before what we never saw before — We had a great battle for skirmishing day before yesterday it commenced only a few hundred yards from where I am writing we made the rebels retreat & yesterday they did now their hand at all — We had gun boats that poured it into them over our heads & threw shells at them — We had a battery on the very spot where I am writing as we moved our camp & tents a mile last evening at sun set nearer where our skirmishing was — We landed at this new place last Tuesday night at mid night — This battery that was where our camp now is also sent shells & lead over our heads & at first wounded some of own men — The mail soon goes & we just had roll call — We started out skirmishing again yesterday morning before 5 o'clock & marched 3 miles over rebel land that up to our arrival here had always been held by the rebels we are now going to take charge of it — I believe in our 1st N.J. Regt. We had meat coffee sugar & crackers given us last night for 2 days — Thee must not be uneasy about me we have an exceeding large force here & we protect one another the rebels have run — I saw 12 dead men yesterday & day before that have died for their country & being as we all are on our side in a good & holy cause they had a happy death — Every man dies when his time comes — There was more killed in 31st & 32nd N.J. Regiments than any others — I hear there are not 10 men left in one co. in 31st that are not killed or wounded — The woods where the skirmishing was I see lived with clothes blankets knapsacks canteens haversacks these small tents &c. Strange it is they were made to carry these into battle — We see blankets coats canteens haversacks & c, of the rebels but not very many of them — We marched over ground yesterday A.M. in the woods where my company saw 7 or 8 dead bodies in one heap —All our skirmishing was in the woods the bullets from rebels whistles & come near my co. but not near enough to hit us — Our capt, when we first went out said all that did not want to go out & fight could step out of our ranks our capt. Said he wanted no man to go if he did not want to But every man went not one of us but are willing to live with & by our flag & die with out it — No one of us wants to be killed but every one of us should be willing to live or die for our country it will be better then for thee, Lucy & the children & children's children of all — We had better be dead than living with rebels in Confederate states — Thee should not at all then mourn after death of a soldier — — By the course of our 1st Lieut. He intends by his course day before yesterday none of us shall be killed because he made some of us come back & not go so near the enemy as the farther we advanced there was the most danger of getting shot — I do not want a public thing made of this so very many would say he

This is what Allison experienced for the first time. Courtesy of the American Battlefield Trust.

was a coward — He is not one our officers want us to kill then but not get killed & wounded in 31st N.J. Regt.

"We have entered the field of battle now in earnest we saw yesterday & day before what we never saw before — We had a great battle for skirmishing day before yesterday it commenced only a few hundred yards from where I am writing we made the rebels retreat" -Joseph Allison

❧

"The troops under Franklin, in transports, reached Yorktown on Monday, 5th inst. Finding the enemy gone and vast quantities of stores of all kinds left on the docks and in the works. A walk through the old town and among the deserted camps and fort convinced me that the rebels had fully expected to hold the place. The fortifications are the finest I have seen in rebellion. Torpedoes, or bomb shells with percussion springs, were sunk in the ground at all points; number suffered from their explosion. We pressed on up York river, reaching this point under escort of gunboats on Tuesday afternoon.

"Deeds of daring heroism were performed. Our men fought well, and the officers behaved gallantly.

"The killed, as far as ascertained, are twenty, and the wounded are about fifty. Troops are pouring in by thousands. Look out for grand events.

"The weather is very fine. Our men are in good spirits and have plenty of good food."

Letter by R. B. Y. to the Editor, *Monmouth County Democrat*, published May 22, 1862, page 1

Addressed to L.A. Allison
Pemberton, N.J.
Let Mother take care of the letters I get from my friends.
   (in pencil) Sunday P.M. May 18 1862

  We left our camp & field of Indian grass this morning & come a few miles & camped — Tell Mother I am (with others) in excellent health & spirits we are confident of success having a glorious & righteous cause — Tell Mother I have stewed some apple since I came here & I eat coffee fresh boiled beef cheese crockery & stewed apple — We come through a beautiful part of the country today we passed by a grist mill the best we have seen in this state & the only one we have seen in operation in Va. I think we will not remain here long orders are continually coming in We come through a rich & beautiful neighbourhood to day which looked like living XXX

❧

Hanover County Va. Saturday May 24, 1862
Dear Sister
  I received thy letter yesterday to day we have rain & cold day we are 12 miles from Richmond our Col (Torbert) read orders to our Regt.

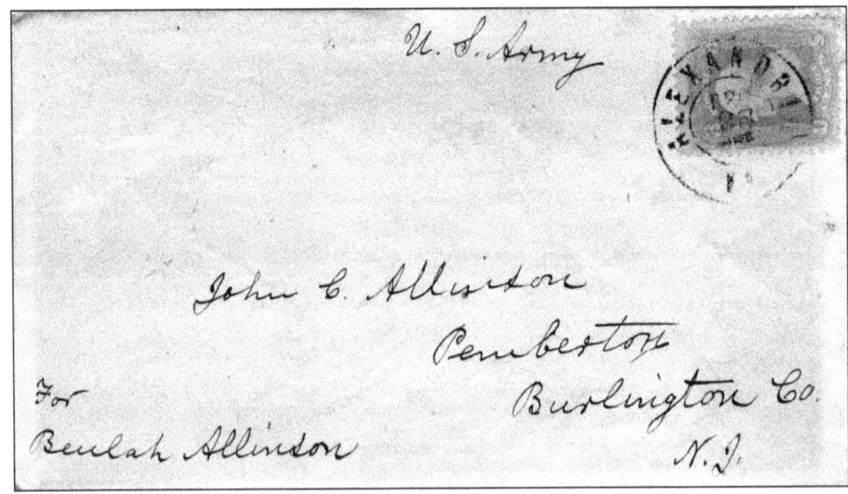

Envelope addressed to John C. Allison, Pemberton Burlington Co. N.J. for Beulah Allison

3 days ago my Co. is on the extreme left & at the time I was on guard so I could not hear the orders — I am informed we are not to write anything home that will benefit the rebels I presume there are rebels in our nobel & distinguished Burlington Co. & our wise rulers think these wolves in sheeps clothing will communicate important information to our enemies here[162] — Tell Mother then I cannot answer all your questions & what I do let it go no further this will be to the advantage of all of us truly in favor of a speedy termination of the war & the reestablishment of one flag, one government & the enforcement of the laws; We may go to day yet & I must close I have not got my Mirror yet of this week. I took supper with our Quartermaster yesterday & a good one it was I saw Capt. Nippins & Capt. Wm E. Bryan yesterday they & their companies are both from Mount Holly-Former is in 4th N. J. latter is in 3rd N. J. Regiments — Liet. King with former has gone home on account of sickness — I gave Mother my address in my last — Willard Wood that thee speaks of is in my Co. he has not been shot at all — He is afraid of it & many of us think he intends not to get shot as he is sick at the time we have sight to get a shot at the rebels I know he has not been with us since we left West Point he got on the sick list & is in hospital or has gone home — I think latter — Wood was a Sergeant — Love to all — We will not remain here long as we often now have a march every day — I got some dried apples of our brigade commissary at 7 ½ cts a pound I find them exceedingly good for my health with the very fat & salt meat we get — I stew them in a tin quart cup I boil I boil my coffee in Tell Mother Sam Read gave me some corn & wheat meal mixed so I am going to have some cakes of my own make — I got some for supper with Read — We have just had coffee & sugar given us for 5 days — Thy brother

    Jos. Allison

❧

The following entries seem to be in the same diary or memorandum form as Allison used before. All are on the same paper and folded together. Did he intend to publish this account in the *New Jersey Mirror*, to which he was a contributor? He does not specifically mention any fighting, although during this time, the 1st Regiment had been present at the siege of Yorktown (April 19-May 4), and the Battle of West Point (May 7-8).

New Kent County Court House
Monday May 12, 1862

The troops on our side are continually pouring in to our aid — Last Friday & Saturday marching here from West Point very near every house of any beauty & respectability had a white flag flying to the breese — Wagon with our provisions did not get here till some time in the night last & today our Col had us to butcher some secesh[163] cattle, we also got some secesh wheat flour, Indian meal, salt, corn, horses, mules, carts & all sorts — Levis in my Co. got a mule, its owner come after it with a flag of truce. Col Torbert let owner have him — Another man brought a white flag & got his mule — through Fortress Monroe also our mail — My friend S. says it is only 18 miles to Richmond — After 6 P.M. & order us for us to move —

May 13. New Kent County Va Bank of Pamonkey river Tuesday

Arrived here at ½ past 1 this morning — Left court house about 9 last evening — The black in their nets catch a great many herring here & a few shad in this river so all our troops today have been living on fish I eat 6 & paid only 10 cts for them & to get them nicely cleaned & fried — Black women sell their corn cakes here very cheap less than half what we formerly paid for them — We have a number of gun Boats in the river very near us — We are here where a bridge crossed this river our side burned it to prevent the escape of the rebels over it — Our brigade & division is here camped upon a very large tract of land of Gen. Lee — The exceeding large field we are in is covered with clover best of pastures — Had a good wash in this river — in battle at West Point last Wednesday the Co. F. & Capt H. Witthack of 31st N.J. Regt. Had a bullet pass through his coat it destroyed one of the buttons on the coat through which the bullet passed — 1St Lieut. is named Van Lier

Pamonkey River Va 14th inst

Rain. On guard at night — Our Regt. Had orders last night to pack up & be ready to march at 3 this morning — We were up at ½ past 2 but only went out on out Picket we took everything with us — Brought some more fresh herring —

Thursday Pamonkey River 15th inst

Rain all last night & rain today — Our Regt. Come in to camp last night from Picket — Got Mirror of last week last night — We expect to move our tents & close up with the Division in which we are in White House Pamonkey river

 Friday May 1th 1862

 Our Regt. Coming in from Picket last night brought a large quantity of straw & corn fodder for our beds — Did not move our tents yesterday as we expected to — All 4 of the first N.J. Regiments are alongside of one another — The number of troops extending here for miles are numerous as the ground is literally all covered with men horses tents cannon & Sutlers in all our Regiments started business again to day in earnest — our sutler sells cheese at 35 cts lb. rasins 50 cts lb. Butter sells with sutler in 2nd N.J. Regt. At 50 cts lb. Jersey penny ginger cakes sell here at 2 cts & 2 ½ cts each — I bought 4 lbs sugar 2 lbs beef & one pound dried apples — 7 ½ cts for last 10 ½ cts lb. for sugar & 8 cts lb. for beef

White House Pamonkey river 17th inst

Clear day. Heard yesterday Richmond is taken by our Union troops — Got orders this morning to pack up & march Got our tents & everything on our backs ready to start (even in the ranks) & the order was countermanded

Sunday May 18th 1862

Took a good wash in this river last night it is fresh water — I got my Mirror last night of last Thursday — I mean last paper published —

Near Kent Co. New Place Monday May 19 1862

Recd orders last night to be packed up ready to march at 4 this morning & advanced towards Richmond some 6 or 8 miles where we reached & camped before 9 this morning — It rained before we got our tents up — This morning on Pamonkey river where we have been for last several days it was 27 miles from Richmond but we in marching have to take those roads where our Infantry Cavalry & Artillery can all march & this takes us not the nearest way by any means — Our Officers all not knowing the roads we often go out of our way — in coming to Slatersville our Major said we had 5 miles to walk some Dunce took us through Barhanesville (this place is spelled as in my last number) where we should not have went at all so we had to walk it is said more than 15 miles — This morning we come alongside of the rail road & telegraph wires part of the time. As we advance towards Richmond the land & country is more valuable & beautiful — It would be strange if it was not so in some degree but there is a vast difference from what it is in the free states — Truly out south here they have gone for cotton & the "peculiar Institution of Slavery"[164] rather than go for something, even at this time would bloom, blossom & flourish more as the rose — We are now only 15 miles or less from Richmond to go return 6 miles we come this morning we will not see 6 fine appropriate suitable houses that in all old Virginia one of the 13 original states, & the state of Washington should adorn & beautify & in some degree, add to & compete with the works of nature. In the last 2 weeks we have visited pine, oak & cedar timber the most valuable — The land is very hilly but there is much land here that is surely rich fertile & productive if rightly man- aged — The other day when we went out on Picket we saw the first field of corn we have seen this year that is up & doing well — They had begun to plough from it as we do in Jersey —

Tuesday 20th inst

Marched a few miles it is near Turnstals Station where we come yesterday but we are not going to Richmond the nearest way — Our Chaplin says the white house is where Washington married

Wednesday 21st inst

Took a long march in warmest part of the day — yesterday we come by a beautiful new white church of moderate sise to day we passed another an old one — Where we camped last night there are a number of houses — There was only one house where we camped near Turnstalls Station[165] — Come to day & camped in a field of Indian grass — Saw very tall pine trees on our road

Thursday 22nd inst

Rain night before last I was on guard 2 hours yesterday A.M. I feel bad for my march yesterday clear last night & this morning it is said we are to march again to day — They were ploughing corn here in a large field where we come yesterday — Ration of whiskey commenced to each & every soldier — some body stole all my sugar so I bought 4 lbs more & 2 lbs dried apples — Eat supper with my friend Read & a good one it was. Sworn in a year ago to day — Capt. loaned me a new Tent because some scoundrel stole half of my Tent — Rain — Recd a letter from Lucy — Wrote to Joe Carr 2 or 3 days ago — Took a good wash — We are in Hanover Co. its last Census is population bond & free 17225 — Same New Kent co. about 6000

Hanover Co. Va. Friday May 23 1862

Warm day — Balloon went up to day & after it gun boats made a charge & rifles & cannon make such a noise we expect every moment to be called on to move — We are just 12 miles from Richmond — Where we were at the White House on Pamonkey river we were on the large tract of land owned by the rebel Lee — This spot is very distinguished in history — Washington owned it & left a home to some blacks for life they do not have any tax to pay I bought some corn meal cakes of them & they fried some fresh herring for me they were very reasonable in their charges — At said White House where we went to from New Kent Co. Court House our chaplin informs me Washington first saw his wife & married her there too — When we come here we met Gen McClellan going the contrary way he rode by our brigade with his hat or cap off (l never saw him with a hat) & near every man cheered him — Read gave me some corn & wheat meal Capt. Loaned me an entirely new Tent — Our Col. commanding us read us an order last Wednesday A.M. before we marched from Gen. McClellan, Gen. Franklin & Gen. Porter that we must not write home any news that will be beneficial to the rebels — I therefore want this kept strictly private.

Hanover co. Va. May 24, 1862

Cold & rain — Had orders in the rain to move but the orders is countermanded so we must remain in this field of Indian grass — Some more rebel prisoners are brought into our Jersey Regiments this is an every day occurrence nearly so I do not often mention it — 26 are brought in to day & day before — These rebel prisoners have gray clothes they all have The most have poor clothes those we see — Sutlers opened for first since we left White House.

Sunday May 25th 1862

Moved again tents & all to near Chicahominee river or swamp or both Come through a beautiful country by some very fine farms & houses also passed a very fine grist mill in operation the only one doing business I have seen in the country in Va. There is a saw mill with this grist mill — We only come a few miles — camped near woods between it & a very fine house & farm in a field of oats.

Near Chicahominee river Monday 26th inst

Got orders to march with 3 days provisions with out tents & knapsacks rain come on & we did not go. Got our 3 days rations in our haversacks

Near Chicahominee river Tuesday May 1862

Very hard rain last night — Sun shines this morning — Recd a letter from cousin Abram & my mirror of last week — Recd a letter from Lucy & wrote to her 2 days ago — Printed order read to us that we are to advance upon the enemy with out our knapsacks thee & our heavy loads are to be packed & left to come with the wagons — This is a wise order

Near Chicahominee Wednesday 28th

Clear warm A.M. rain P.M. — No news — Our brigade commissary sells potatoes at 90 cts a bushel he did sell them at 80 cts — Tried to get some more dried apples commissary has none now — A fine horse at work for Read was so bad he was shot by a man from Camden this was done to put an end to his sufferings.

Near Chicahominy Thursday 29th inst

Clear day, More rebel prisoners taken with them a rebel general.

New York papers & Herald Times & Tribune are selling at 10 cts each — Baltimore Clipper at 5 Cts a copy — Inspection every day — Nothing new, been here in camp several days

Chicahominy 30th inst

Regt. Detailed to cut timber for bridges I was sent to unload wagons to the beautiful & very valuable residence of Doctor Wm F. Gains this man has 13 hundred slaves — To day fine barn of W. F. G. turned into a hospital for wounded soldiers of rebel army now in our hands — I helped to take 21 out of our ambulances — most of them are from North Carolina & are very badly wounded — 2 have each lost their left leg (one below other below knee) one or two lost one arm & one lost his eye sight caused by one of our shells — 2 of these 21 come from Lousiana & one of them from New Orleans — one is a boy 18 years Old from old North state is shot in back as other are — Bodine sharp shooters come to protect these wounded there is 2 Doctors our of rebel army come with a flag of truce to take care of these their wounded men. One is very kind other does not appear to care at all about their welfare — Bodine[166] sharp shooters have a calf skin (with hair on) on their knapsacks — Exceeding heavy rain P.M. much thunder and lightening — More wounded men com in

Enslaved people on the William Gains plantation, Hanover County, VA; George Harper Houghton photograph, courtesy of Library of Congress.

while I was unloading wagons I left 36 & more were coming — Had
been wounded I5 miles from this point when they were brought —
Wounded a week tomorrow.

*". . . the beautiful & very valuable residence of Doctor Wm F. Gains this
man has 13 hundred slaves — To day fine barn of W. F. G. turned into a
hospital for wounded soldiers of rebel army now in our hands"*
— Joseph Allison

Saturday in Camp 315t inst

Very heavy rain last night — Distance to Richmond 8 miles
Chicahominy ½ mile I mean from where I was yesterday — Heard
drum beat of enemy this morning — yesterday could see smoke of
rebel cannon as they & we have both been firing away their shells come
towards us & burst but have not yet hurt us — There is considerable
tobacco that has been raised here — Mailed a letter to Father of my
first Lieut. Appearance of rain — Our soldiers get tobacco out of the
barns — Our camp is near Chicahominy

Candlelights

There has been a hard fought battle to day or mean in the afternoon it was continued till after dark we heard the cannons rifles &c in
such quick constant & rapid succession there must have been bloody
work — I hope our side has got the advantage we will hear tomorrow

Chicahominy June 1st 1862

At midnight last night our cooks were got up to cook for our Co.
so that it could march by day light or before our Regt. Was up in night
to pack &c but we are not on advance yet — I am detailed to carry
water &c for cooks — A little rain — Battle going on; we can hear the
musketry quite plain — Have inspection every day — Cleared off &
very hot day — now candle light & we did not move=Great deal of
cannonading done to day on our side, we could see the rebels run on
their horses — 2 men in our Co. died since 15 of last Mo in hospital.

"Battle going on; we can hear the musketry quite plain — very hot day
— we did not move — Great deal of cannonading done to day on our
side, we could see the rebels run on their horses" — Joseph Allison

The Monmouth Democrat newspaper[167] carried extensive war news coverage, including day-to-day summaries of Gen. Franklin's efforts to push toward Richmond. The May 15, 1862, issue began by stating "McClellen is doing wonders on the Penninsula, driving the rebels from all points . . . Before the week is out he will in all probability plant the Stars and Stripes in that city."

The Battle of Gaines' Mill, sometimes known as the Battle of Chickahominy River, took place on June 27, 1862, in Hanover County, Virginia, as the third of the Seven Days Battles of the Peninsula Campaign. There were almost 2,000 men killed, including several of Allison's friends. Thousands more suffered wounds or went missing. This was a tactical defeat for the Army of the Potomac.

This is the last of Allison's letters in the collection. He was wounded at Manassas on August 30, 1862.

Many Burlington County families anxiously awaited the regular newspaper reports such as the *New Jersey Mirror*, edited by Joseph Allison's friend Joe Carr, to know the fate of loved ones. Readers discovered this terrible news on July 10, 1862:[168]

The accounts of the series of battles in front of Richmond, give glorious yet melancholy evidence of the valor and efficiency of the officers and men of the Jersey Regiments. They were in the thickest of the fight, and bore themselves like true sons of their brave sires. Though forced back by superior numbers, they fought to the last— bravely, steady, stubbornly—and no stain rests upon their banner. They were brave men. They fought, they died, like heroes. Many who went from our own County, head the roll of honor, which will be transmitted to posterity, containing the heroic names of Jerseymen returned as "Dead on the field of battle." A long and sad list it will be, when completed; and the pangs which will crush the hearts of bereaved friends, widows and orphans, will be a part of the price which our loyal State pays for the preservation of the Constitution and the Union. In times of peace and prosperity, we are too apt to forget the value of good Government; and it is only when such incidents as those to which we refer, bring painfully to our minds the value of our free institutions, that we properly appreciate them. We may weep for those who have fallen in the fearful strife, but could we wish them to die a nobler death? The pangs which lacerate the hearts of loved ones now, are but temporary, but the glory won by deeds of generous daring and devoted patriotism, will live while our country's annals endure.

THE KILLED, MISSING AND WOUNDED. The Fourth Regiment suffered severely.—At one time they were completely surrounded by the enemy, and were obliged to fight their way out, leaving a noble band, dead and dying, on the field of their terrible encounter. According to the statement of an officer, this regiment went into fight with 650 men, and only 75 reported themselves afterwards. Six of the captains were killed in an hour after the battle commenced.—Among them it is feared is Captain William Nippins, of this town (Mount Holly). Later accounts state that Col. Simpson, of the Fourth Regiment, who was reported killed, is a prisoner, known to be in Richmond, with about 450 of the Regiment. We trust this may prove true. Col. Simpson is a native of Burlington County.

CAPTAIN NIPPINS' COMPANY. The following is a list of the Missing in Captain William Nippins' company of this town (Mount Holly). This company is in the Fourth Regiment. The list was brought to us by Isaac T. Ager, who left camp on Friday. All that is positively known of them, is that they went into the fight, and nothing was heard of them afterward. Some of them are doubtless killed, and the rest are probably prisoners—many of them, most likely, being wounded. We learn that William Bodine of this town (Mount Holly), lost an arm, and was in the hospital at the time the enemy took possession of it: OFFICERS: William Nippins, Captain, John L. Ridgway, 1st Lieutenant, James H. Brewin, Sergeant, Charles H. Glenn, do., Benjamin F. Stidfole, do., Samuel D. Cross, do., Caleb M. Wright, do., Leander Brewin, do., John W. Hooper, do., Charles Hill, do., Charles P. Cline, do., Joseph Young, do. PRIVATES: Joseph Boxer, Joseph Broom, Joshua Bishop, George Clevenger, John Cithcart, Cornelius Donaver, James N. Garle, Charles Groome, Charles B. Hodson, Woodrow Hughes, Stephen Hoffman, Charles Johnson, John Kirby, Job Kinsinger, Charles S. Moore, John Muckrey, John S. Moore, Thomas Mathis, Daniel Nixon, Augustus Pittman, Robert Roff, Job Stockton, Peter S. Shemelia, Joseph Scroggy, William

Bodine, John Brown, Ambrose Copp, William Clevenger, Stephen Cithcart, Theodore Fowler, Thomas Green, James Giberson, Rheese Horner, Andrew High, Jacob Hoffman, James Kain, Joseph King, Peter Leonard, John McClarney, Walter Mitchell, John H. Morton, Elwood Nixon, John W. Price, George C. D. Powell, Henry Rossel, John O. Shemelia, Michael Sorceny, Adam Stiltz, Jackson Quigley. The above are nearly all of Capt. Nippins' Company that were engaged in the fight—there being about eighty men in the ranks when the conflict commenced.

CAPTAIN BRYAN'S COMPANY. We have also received from Capt. William E. Bryan, of this town—through Mr. Ager—the following list of killed, wounded and missing, in his company. This company is in the Third Regiment: KILLED: William Scott, James Farley, Charles Read, John Ribad, Walter Mudford, James Bellsher. WOUNDED: Sergeant Devinney, Michael Hogan, Charles Bennett, William Glenn, Michael Mick, Charles Down, Charles Dennis, Thomas Dennis, John R. Middleton, Lemuel Middleton, Samuel Thompson, Hezekiah Peterson. MISSING—Garvin Neilson.

CAPTAIN BUCKLEY'S COMPANY. The following is a list of killed, wounded and missing, in Capt. Buckley's Company, from Burlington, formerly Capt. Rowand's. Capt. Buckley is said to be a prisoner: KILLED: Corporal Charles Derey, Watson H. Miller, John Rogan. WOUNDED: Sergeant Steward, Corporal B. Riley, John

Perhaps this field hospital treated Allison when he "had his leg shot off."

Adams, Thomas F. Dunbar, Timothy Titus. MISSING: Theodore Dare, William Benham, Thomas S. Palmer, John Andrews, Stephen Butler, Capt. D. P. Buckley.

CAPTAIN HALL'S COMPANY: Capt. Hall's Company, from Moorestown, is in the Fourth Regiment. No information has yet been received of its loss.

New Jersey Mirror , September 4, 1862: "Joseph Allison of this vicinity, we regret to say, had his leg shot off in the recent battles."

New Jersey Mirror, September 18, 1862, page 3: "I had almost forgotten to say I have in my ward Joseph Allinson, known as the Springfield orator. His leg was amputated about two weeks since; he is doing very well; and would do better if he was more obedient; but do as will, coax, threaten, or scold, it is always the same; his knowledge is equal if not superior to my own, and his own way he will have; he is exceedingly loquacious and indulges the hope of getting back to Burlington County and having a bout with the great men."[169]

On the back of one of the envelopes in the collection is a penciled comment "killed."

The *New Jersey Mirror* October 9, 1862, carried this simple obituary: "In Washington, D. C., on the 2d instant, Joseph Allinson, of the 1st Regiment New Jersey Volunteers, formerly of Burlington County, in the 37th year of his age."[170]

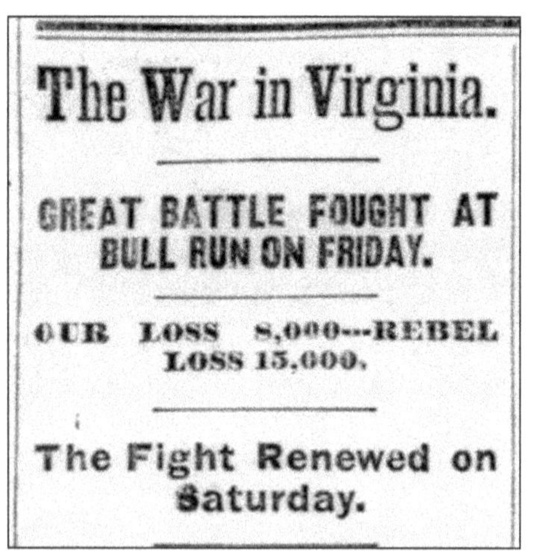

APPENDIX A
LIST OF MEN MENTIONED IN JOSEPH ALLISON'S CORRESPONDENCE

This is an attempt to honor those who served with Allison by determining their service record, and what happened to them. Many of them have descendants still in New Jersey. Joseph Allison's letters provide a first-hand account of contact with them during the war. The National Park Service List of Soldiers and Sailors[171] yielded the information provided below along with the New Jersey Department of State Civil War Records,[172] Newspapers.com, and public records found through Ancestry.com. The information below represents the sum total of information about these men found in these sources.

Isaac T. Agar (1821–1886) apparently had not yet enlisted when he met Allison at camp. Agar was commissioned as 2nd Lieut. in the 34th Infantry, Co. B from October 27, 1863, to May 5, 1865. He returned home to Burlington, where he worked as a plumber.

PVT Joseph Allison (1826–1862) 1st Reg. Infantry, NJ Volunteers, died from complications of wounds suffered at the 2nd Battle of Bull Run.

CPL Thomas B. Arey probably was related to Joseph Allison through Allison's mother. They got together while stationed at Camp Seminary to exchange some clothes. Arey was listed as "KILLED" in the July 24, 1862, *New Jersey Mirror* article of the latest casualties from Captain Buckley's Company (3rd Regiment, Co. C) in action at Gaines Farm, Virginia., June 27, 1862.[173]

CPL Winchester T. Bennet—Private, 1st Regiment, Co. D., killed in action at Gaines Farm, Virginia., June 8, 1862.

CAPT William E. Bryan (1822–1893) was a tailor from Northampton when he enlisted in the 3rd Infantry in June 1861. He had become a

Captain in Company H of the 3rd Regiment when Allison met up with him in Hanover County, Virginia in May 1862. Later, he was promoted to Major and mustered out in June 1865. He is buried in the Mount Holly Cemetery.

LT William H. Campion In the 1860 federal decennial census, he is still a medical student living with his family in Mount Holly. He visited Allison in the field.

PVT Theodore Carhart (1838–1883) shared a tent with Allison at Camp Seminary and complains about the postmaster charging postage on letters he wrote home. He was promoted to Corporal in August of that year and mustered out in June 1863. Like several others in the 1st Infantry Regiment, he was from Phillipsburg.

PVT William Cox (1840–1914) was a shoemaker from Pemberton. He suffered a shot in the arm while on duty around Fairfax, Virginia, and bravely refused any anesthetic during the surgery to remove the bullet from the bone.

PVT Barnet (Barney) Devlin, who had been shot in the finger around Fairfax, Virginia, was killed in action at Gaines Farm, Virginia., June 27, 1862.

COL Elmer Ellsworth (1837–1861) was the first Union officer to be killed in battle. He was a lawyer and close friend of President Lincoln. Allison was very moved by seeing the place where he was killed.

PVT George Emmons, served a 3-year enlistment in the 1st infantry Regiment, Co. D.

PVT James Flood had been a tentmate of Allison's at Camp Seminary. Allison reported his death to the *New Jersey Mirror* when Flood was killed in action at Gaines Farm, Virginia, June 27, 1862.

Gen. William B. Franklin (1823–1903) led the Peninsula Campaign. He was a civil engineer both before and after the war. Allison served in Franklin's brigade.

PVT John Gano was killed in action in the same battle as James Flood, at Gaines Farm, Virginia., June 27, 1862. Allison saw him fall.

CAP Joseph Gale was the recruitment officer when Allison enlisted in the New Jersey Militia. Gale served out his 3-month enlistment.

PVT Benjamin Gaskill (1838–1916) was from Springfield Township. In 1861, he enlisted for 3 years in Co. C, 1st Cavalry Regiment. After the war, he returned to his farm. He is buried in the Mount Holly Cemetery.

Dr. Charles C. Gordon was the physician from whom Allison tried to get an excuse to get out of guard duty at Camp Seminary. He was Assistant Surgeon for the 1st Infantry, and was mustered out as Supernumerary, February 16, 1863.

MAJ David Hatfield (1831–1862) served in 1st Regiment, Company A at the Fairfax encampment when Allison went to him for some kind of permission which he was unable to give. Prior to the war, he had been a night watchman in Elizabeth, New Jersey. He died at Elizabeth July 30, 1862, of wounds received in action at Gaines Farm, Virginia, leaving a wife and 4 children.

William H. Johnson – Co. B, 1st NY Cavalry, shot for desertion Dec. 13, 1861, at Fairfax, Virginia. It was the first execution of the Army of the Potomac. Allison was among the troops who witnessed the execution. Several accounts and illustrations can be found online.

SGT Martin VanBuren Hargrove (1839–1893) from Pemberton was the 1st Sergeant of the 23rd Infantry, in a 9-month enlistment that ended in June 1863. He added a note to the back of the letter received by Antis Robbins, telling her about her husband. After the war, he worked as a clerk in a store.

LT John L. King had enlisted as a private and had been promoted by the time the N.J. Volunteers reached Hanover County, Virginia. He became ill, returned home, and was mustered out.

PVT Joseph A. Lamb (1841–1893) was one of Allison's traveling companions on a pre-war trip to Ohio. He served with the 23rd Regiment N.J. Volunteers, Co. D and was wounded at the Battle of Salem Church on May 3, 1862. He mustered out in June 1863 and returned to his farm in Springfield, New Jersey.

PVT Joseph Levis, 1st infantry Regiment, Company D, got a mule at New Kent Court House but had to give it back when the owner showed up with a truce flag. Levis mustered out in September 1864.

COL William R. Montgomery was 1st Infantry Colonel at Camp Seminary, then was promoted to Brig. Gen., U. S. Vols.

CAPT Valentine Mutchler (1828–1877) was Allison's commanding officer in the Fairfax, Virginia, encampment. He was a Captain in Company D of the 1st Regiment, was promoted to Major in order to fill a vacancy and resigned in 1863. He had been a contractor in Phillipsburg before the war and returned there upon leaving the service.

CAPT William Nippins (1828–1865) of Mt. Holly met Allison in Hanover County, Virginia. Before the war, he was a salesman living in Northampton. He was in Company D of the 34th Regiment. His death is recorded in the U.S. Register of Deaths of Volunteers, and he is buried in St. Andrews Cemetery, Mount Holly.

Charles Norcross (1831–?) and his brother Earl were both in the 4th New Jersey Regiment.

PVT Earl Norcross, Co. I, 4th New Jersey Regiment. PVT Norcross died of typhoid fever at the U.S. Army General Hospital at Camp Seminary on February 2, 1862.

Gen. Andrew Porter (1820–1872) had fought in the Mexican American War. He was a staff officer of Gen. McClellan during the Peninsula Campaign.

Sam Read was the 1st Infantry Quartermaster, whom Allison mentioned several times during his assignment to Fairfax, Virginia. He mustered out in 1864. There were several men by that name, but this one may have been from Elizabeth, New Jersey.

Aaron S. Robbins (1818–1909) While not mentioned in Allison's letters, his story is part of this book. At age 43, Robbins enlisted in the New Jersey Militia, then was commissioned as a 2nd LT in the 1st New Jersey Cavalry. When that enlistment ended, he joined the Twenty-Third Regiment (the Yahoos) and was taken prisoner with John Allinson at the Battle of Salem Oak. Robbins returned home, and worked as a farmer, storekeeper, and postmaster. He is a great-great grandfather of the author.

PVT James E. Ross, team driver, 1st Infantry Regiment, Co. D. Ross bought a watch from Allison for $16, a $3 profit over what Allison had paid for it. When his 3-year enlistment was up, Ross re-enlisted on December 28, 1863, and served in Co. A. 1st Battalion. After the war he moved to Beverly.

CAPT Joseph French Rowand (1833–1862) 3rd Regiment, Co. C. was a carpenter from Mount Holly. He resigned on Jan. 20, 1862, returned home, and died a month later. Allison was much grieved by his death.

PVT Jacob R. Seeds was still a civilian in Burlington County when he wanted to subscribe to the *Tribune*. He enlisted in the 2nd Cavalry in 1863, and was accidentally killed at White's Station, Tennessee, Aug. 5, 1864.

PVT John Schoonover (1840–1917), a Sussex County farmer was the 1st Regiment Captain's clerk at Camp Seminary who helped Allison get out of the Provost Guard House. He was promoted to Adj. August 2, 1862; then to Lieut. Col. and became Bvt.[174] Col. March 13, 1865.

Joseph Arey Shinn (1843–1918) was the brother of Vashti Shinn and the recipient of several letters from William. He served in Co. B. 3rd NJ Regiment as a wagoner from May 25, 1861, to June 23, 1864. Following the war, he returned to farming.

PVT William Shrope (1821–1887) began as a wagoner for 1st Infantry, Company D, then became a soldier. He was discharged for disability from Camp Seminary in 1862. He returned home to Warren County, where he worked as a laborer.

Col Simpson (This could be either Benjamin F. or James H., both were in the 4th Regiment) The *New Jersey Mirror*[175] reported that Col. Simpson of the Fourth Regiment, who was reported killed, is a prisoner, known to be in Richmond.

CAPT Charles Sitgreaves Jr. (1844–1899) of the 1st Infantry, Company D, was promoted into Mutchler's place. He mustered out in June 1864. Following the war, he returned home to Phillipsburg, where he managed the family's real estate holdings.

Charles Sitgreaves Sr. (1803–1878) was mayor of Phillipsburg, then served 2 terms in the New Jersey legislature as a member of the Democratic party. In 1862 it was suggested by his hometown newspaper that he be nominated for Governor of New Jersey, which Allison supported. Sitgreaves was a banker, with investments in railroads.

PVT Samuel Smith was still a civilian when he subscribed to the *Weekly Tribune* newspaper that Allison distributed. He enlisted in the 1st Cavalry in September 1864.

CPL John D. Thompson (1830–????), 28th Regiment, Company I, enlisted for a 9-month term, and mustered out in July 1863. He was from Burlington.

COL Alfred T. A. Torbert (1833–1880) was a graduate of the United States Military Academy, and a career military man. By age 20, he was a seasoned combat veteran. Allison's letters were written when Torbert was commanding officer of the 1st NJ Regiment. In November 1862, at age 29, he was promoted to Brigadier General. Following the war, he held several diplomatic posts. He died in a shipwreck off Mexico.

COL. Isaac M. Tucker took over 1st Infantry, Co. D., from Col. Montgomery at Camp Seminary. He was killed in action at Gaines Farm, Virginia., June 27, 1862.

LT John P. VanLeer (??–1862) was killed in action near the Pamunkey River on May 5, 1862. Allison reported seeing a bullet pass through his coat.

COL J. W. Wall, Allison intended to write to him. May have been James Wall, but unable to find a Colonel by that name.

SGT Washington W. Watts (1841–1912) was one of the men with whom Allison went into Alexandria one day. After the war he became a railroad fireman and lived in Pemberton.

Dr. Edward L. Welling (1835–1887) from Pennington was a 26-year-old assistant surgeon to the 3rd Regiment when he treated Joseph Allison at the hospital on the Virginia Seminary grounds. After mustering out in June 1865, he returned to private practice in Mercer County, where he and his wife raised their family. He lived to be 62 years old.

CAP H. Witthack of the 31st Infantry Regiment had a bullet pass through his coat and destroy the button during the battle at Pamunkey River.

SGT Willard S. Wood, 1st Infantry, Co. D., was a friend of Allison's mother. On April 29, 1862, he was ordered to collect descriptions of the men so they could be identified if they became lost. He was killed at Spotsylvania Court House in May 1864.

SGT Jethro B. Woodward, (1826–1908) was the Quartermaster for Company D; he was with Allison when they went to capture some wood from the rebels on the railroad from Alexandria to Richmond. He mustered out on August 1, 1863. After the war, he moved to Tioga County, Pennsylvania.

126

APPENDIX B
The *New Jersey Mirror* reported these deaths of men in the 23d Regiment, New Jersey Volunteers

Burlington County paid a heavy price during the war. These are the deaths, copied exactly as they appeared in the newspaper, of members of the Twenty-Third. The oldest soldier listed here was 45 years old when he died; the youngest was 16.

In camp, in King George county, Va., on the 12th December 1862, in the 20th year of his age, Harrison DeCamp, company D, 23d Regiment, New Jersey Volunteers—late of Cookstown.

In camp, in King George county, Va., on the 14th December 1862, in the 40th year of his age, Christian Spear, company B, 23d Regiment, N. J. V. (New Jersey Volunteers)—late of Bordentown.

In camp, near White Oak Church, Va., on the 24th December 1862, in the 19th year of his age, George H. Duble, company G, 23d Regiment, N. J. V. (New Jersey Volunteers)—late of Mount Holly.

In camp, near White Oak Church, Va., on the 25th December 1862, Aaron Ridgway, company E, 23d Regiment N. J. V. (New Jersey Volunteers)—late of Jobstown.

December 25, 1862, We learn that Gen. Torbert's Brigade, consisting of the 1st, 2d, 3d, 4th, 15th and 23d, has lost: killed, 13; wounded 83; and missing, 69.

George A. Gilbert, of Company I, (Capt. Burnett's) 23d New Jersey Regiment, died at the Camp of the First N. J. Brigade, King George County, Va., on the 7th December 1862, of Typhoid Pneumonia. Mr. Gilbert was a printer, a native of Burlington, and formerly worked in the office of the *Mount Holly Herald*. He was in the 45th year of his age.

On the 10th April 1863, in Regimental Hospital, near White Oak Church, Va., Samuel C. Schooley, of 23d N. J. Regiment, son of Asa H., and the late Phebe (sic.) Ann Schooley, of Moorestown.

In Regimental Hospital, near White Oak Church, Va., April 10th, George Goodman, Co. D., 23d N. J. Volunteers.

In the same hospital (Regimental Hospital, near White Oak Church, Va.), on the same day (April 10th), Edwin M. Hendrickson, Co. H 23d N. J. V. (New Jersey Volunteers).

In the same hospital (Regimental Hospital, near White Oak Church, Va.), on the same day (April 10th), Charles Schooley, Co. H. 23d N. J. V. (New Jersey Volunteers).

In the same hospital (Regimental Hospital, near White Oak Church, Va.), April 11th, John Mullen, Co. I. 23d N. J. V. (New Jersey Volunteers).

In the same hospital (Regimental Hospital, near White Oak Church, Va.), April 14th, Daniel Read, Co. D. 23d N. J. V. (New Jersey Volunteers).

In the same hospital (Regimental Hospital, near White Oak Church, Va.), April 16th, James Thomas, Co. K. 23d N. J. V.

In Camp near White Oak Church, Va., on the 5th March 1863, John W. Boyd, of Capt. Newbold's company D., 23d Regiment N. J. V. (New Jersey Volunteers), aged 27 years.

In the same camp (as John Adams, near White Oak Church, Va.), on the 13th March 1863, Daniel McIntire, of the same company (Co. B., 23d Regiment New Jersey Volunteers), aged 38 years.

In the same camp (as John Adams, near White Oak Church, Va.) on the 14th March 1863, James G. Ivins, of Co. L., 23d Regiment N. J. V. (New Jersey Volunteers), aged 23 yrs.

In the same camp (as John Adams, near White Oak Church, Va.), on the 14th March, 1863, Henry Larzelere, of Co. A. 23d N. J. V. (New Jersey Volunteers).

At White Oak Church Hospital, Va., on the 7th March 1863, Mark J. Rines, of Co. A, 23d Regiment, N. J. Volunteers, aged about 16 years.

In Regimental Hospital at White Oak Church, on the 19th February 1863, of typhus fever, Joshua L. Joyce, of Co. F. 23d Regiment N. J. V. (New Jersey Volunteers), eldest son of Allen V. and Susan E. Joyce, in the 20th year of his age. "His battles are fought, / And his march—it is ended: / The roll of the drum / Shall wake him no more."

On Sunday, the 15th February 1863, near White Oak Church, Va., Corporal Charles Sharp Heisler, of Co. G. 23d N. J. V. (New Jersey Volunteers), son of Geo. M. Heisler, of Beverly, aged 24 years.

In camp, near White Oak Church, Va., on the 10th February 1863, Isaac Wells of Co. H. 33d N. J. V. (New Jersey Volunteers), aged 18 years.

In camp near White Oak Church, Va., on the 13th February 1863, Firman Rambo, of Co. C. 23d N. J. V. (New Jersey Volunteers), aged 24 years.

In Hospital, at White Oak Church, Va., on the 6th of February, Amos S. Austin, of Co. F. 23d Regiment N. J. V. (New Jersey Volunteers).

In camp, near White Oak Church, Va., on the 25th December 1862, Aaron Ridgway, company E, 23d Regiment N. J. V. (New Jersey Volunteers)—late of Jobstown.

In White Oak Church Hospital, Virginia, on the 24th December 1862, George Duble, son of Isaiah C. Duble, of this vicinity, aged 22 years. Deceased was a member of Capt. Ridgway's Company, Burlington County Regiment. He was in the battle of Fredericksburg and escaped unhurt. At the close of the fight, feeling unwell, he was sent to the hospital, where he remained until his death.

A NATIVE OF BURLINGTON KILLED IN BATTLE. January 22, 1863, The following notice of death of Richard W. Chase, a native of Burlington, we copy from *The Philadelphia Inquirer*: RICHARD WYATT CHASE. Among those of the "noble three hundred," as Gen. Stanley calls them, of the Anderson Troop, who fell on the bloody field of Murfreesboro, fighting for their country and for civilization, no one will be more missed from a wide circle, nor more deeply mourned, than the young soldier whose name heads this brief notice. The two brothers Chase—Richard Wyatt and William Beverly—joined the Anderson Troop as privates. The first was killed, the last 'missing,' whether killed or a prisoner, is not known. These young gentlemen were grandsons of the late Bartholomew Wistar, a prominent and influential Quaker of Philadelphia.—They were connected with many of our best families. Richard, of whom, from the melancholy certainty of his fate, it is proper to speak, was distinguished by signal virtues. In him, intellect and education were united to many graces of person. Few young men combined so much mental and moral power as he. He had just attained his majority, and an important position with the firm of I. P. Morris, Towne & Co. The latter he left to join the army from a sense of duty. Full of courage and hope and faith, he behaved, in the field, gallantly, as became him. On Monday, the 29th of December (1862), whilst riding beside his brother, in fine spirits, previous to the action, he recited portions of Tennyson's "Princess," and the following detached extracts from Morte d'Arthur: "I have lived my life, and that which I have done / May He, within himself, make pure! but thou, / If thou shouldst never see my face again, / Pray for my soul. More things are wrought by prayer / Than this world dreams of." Still riding he continued: "Now I go / To the island valley of Avillon, / Where falls not hail, or rain, or any anew, / Nor ever wind blows loudly; but it lies / Deep meadowed, happy, fair with orchard lawns, / And bowery hollows, crowned with summer seas. / There I will heal me of my grievous wound." Almost with the immortal words warm upon his lips, he received a shot in the head, that must have been fatal on the instant.—Thus he fell, adding another to that innumerable company happily described by Kossuth as "the unknown demi-gods who die for country and for man." Since the above was written, a letter has been received from the brother of the dec'd.—W.

Beverly Chase, stating that while he was standing by Richard's body, a party of Rebels took him prisoner, robbed him of all his valuables, and paroled him on the spot. He is now in Nashville, and expects to come home soon.

Inscription on the White Oak Church Historical Marker, Fredericksburg, Virginia

Across the road stands White Oak Church, an important Civil War landmark during the winter of 1862–1863. Stafford County Baptist constructed the simple weatherboard structure sometime after 1789, later adding an attached shed with a separate entrance for African-American members of the congregation. A Union soldier described it disparagingly as a "miserable, insignificant structure, dilapidated and steepleless, and seems to have belonged to some former age. It looks," he thought, "very much like some ancient horse shed and barn that may be seen in some of our less thriving villages."

With the arrival of the Army of the Potomac in November 1862, White Oak Church instantly became the center of one of the largest communities in Virginia. For seven months, 20,000 soldiers of the VI Corps camped in the immediate area. During that time the church served alternately as a military hospital, a United States Christian Commission station, and as a photographic studio. Fifty-two soldiers who died during the encampment were buried on the church grounds. Their bodies were later moved to Fredericksburg.[176]

APPENDIX C
LIST OF MEMBERS OF BURLINGTON PREPARATIVE MEETING
Prepared January 12, 1805, approved by William Allinson

Explanatory Notes &c	Members Names	Births or when rec'd either by Applicat. or Certificate	Married
Wife of the ? Allinson	Sarah Allinson		
Daughter of Isaac & Sarah Cooper of Woodbury ?	Martha Allinson, Widow of Sam'l Allinson	Came last to Burl'n by Certif't from Evesham 5th 11 mo. 1795	
	his Children		
Wm. & Mary are Children of afores'd Allinson & his wife Elizabeth, the Daughter of Ab'm & Eliz'th Smith of Burlington	William Allinson, Came last by Certif't from Evesham rec'd 6th 11 mo. 1794	born at Burl't 11 mo. 15. 1766	
	Mary Allinson, Came last by Certif't from Evesham rec'd 5th 11 mo. 1795	born at Burl't 2 mo. 9. 1772	
	David Allinson, Came last by Certif't from Phila. rec'd 5 mo. 1808	born at Burl't 1 mo. 14. 1774	
	Elizabeth Allinson, Came last by Certif't from Evesham rec'd 5th 11 mo. 1795	born at Burl't 7 mo. 26. 1775	
	Sibyl Allinson, Certif't rec'd last from Evesham 5th 11 mo. 1795	born at Burl't 12 mo. 26. 1779	
	Margaret Allinson, Certificate rec'd from Evesham 5th 11 mo. 1795	born in Wales 3 mo. 12 mo. 1781	Mar'd 3 mo. 24. 1831 to Benj'n Fairbanks
	Samuel Allinson, Certif't rec'd from Evesham 9th 11 mo. 1797	born in Wales 5 mo. 7. 1784	Mar'd 4 mo. to Sus'a Deliv'h Smith
	John Cooper Allinson, Certif't rec'd from Evesham 5th 11 mo. 1795	born in Wales 7 mo. 25. 1781	

APPENDIX D
ALLISONS IN BURLINGTON MONTHLY MEETING
BURYING GROUND

Provided by the Burlington New Jersey Quaker Meeting House and Center for Conference.

List does not include Allinson daughters who are buried under their married surname. Many in-laws rest here also. It is unclear whether there are actual remains in all these plots, or if some were reserved for the families named on the diagram.

John Cooper Allinson
Susan Allinson
Rebecca W. Allinson – wife of J.A.
Lucy Allinson – wife of J.C.
William James Allinson
William Allinson – son of W.J.A.
Sybil Allinson
Elizabeth Allinson
Edward Allinson
Mary Allinson
David Allinson
Thomas Allinson

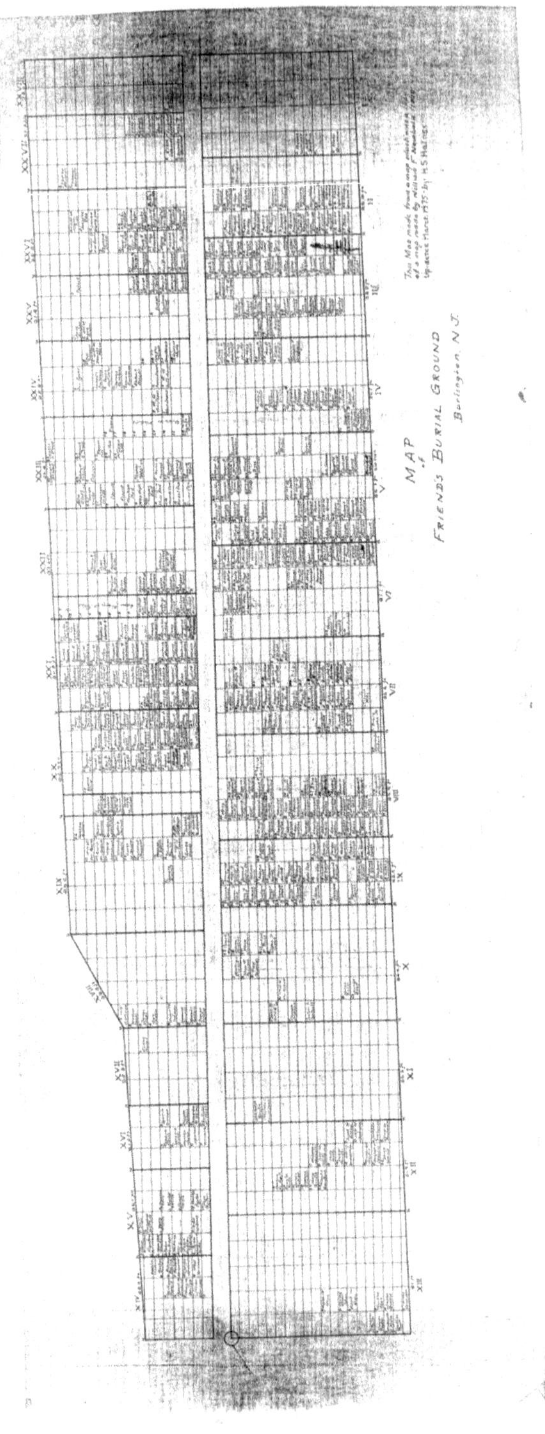
MAP
of
FRIENDS BURIAL GROUND
Burlington N.J.

APPENDIX E
GLOSSARY OF QUAKER TERMS[177]

The Religious Society of Friends – A religious group that originated in England during the middle of the seventeenth century of individuals who dissented from the Anglican Church that emphasized ceremony or creed and turned inward in search of a personal experience and direct communication with God. This is a faith-based community of followers who recognize God's light in every person.
•   Friends – the term by which followers refer to themselves and each other; also used as a descriptor, such as Friends' Meeting, Friends' School or Friends' Burying Ground
•   Quaker – originally a derogatory term that has become an acceptable descriptor that others use when referring to Friends or their institutions, customs, and beliefs

Philadelphia Yearly Meeting – the central organizing body for Friends' meetings in the Philadelphia, Pennsylvania, United States, area, including parts of Pennsylvania, Maryland, Delaware, and New Jersey.

Monthly Meeting – the basic administrative unit for each congregation comprising members of the Religious Society of Friends. It is here that Friends conduct regular worship service, known as "meeting for worship."

Quaker meeting is a faith community first and foremost. People who visit or associate with a Quaker community gather for worship, to conduct business, share fellowship, and to manifest Quaker leadings in the world. These gatherings are called meetings.

Meeting for Worship – Friends celebrate unprogrammed worship; all members are ministers in one way or another. For this reason, there is no paid clergy and no pre-arranged liturgies for worship. Those congregant meetings under the Philadelphia Yearly Meeting practice silent worship, in which the members sit in contemplative silence and prayer until the Spirit moves a member to stand and make a declarative statement meant to edify those present.

Minutes – the written record of concerns, decisions, and agreed upon actions, based on consensus, taken by the meeting membership. A guide to understanding the abbreviations used in Quaker minutes is found at A Friendly Glossary :: Friends Historical Library :: Swarthmore College.

Quaker meeting house – common term for any building in which Quakers gather. At present, there are more than twenty meeting houses in Burlington County, New Jersey. Some of these are active, others have been "laid down" (closed).

Clearness Committee – A clearness committee is a group of Friends appointed to aid the meeting as a whole, and individual meeting members, find focus and clarity around concerns, decisions, or actions.[178]

Burlington Quaker Meeting House & Center for Conference, 340 High St., Burlington, N.J. 08016. In 1991, the Burlington Monthly Meeting was laid down and the Meeting House closed. It has since been restored and opened as a conference center behind it for the use of PYM Quaker youth and families and other Quaker and nonprofit organizations.[179] The Center administers and maintains the Burlington Friends Burying Ground behind it.

Testimony – what Quakers call the ways found to live and act, based on beliefs.

Weighty Quaker/weighty Friend – An informal term for a Friend who is respected for spiritual depth, wisdom, and long service to the Religious Society of Friends.

**APPENDIX F
LETTERS AND MEMORANDA INSTALLMENTS
FROM JOSEPH ALLISON**

Date   Location
29 April 1853   Baltimore
3 May 1858   Camden
5 September 1861   Near Theological Seminary, Virginia
8 September 1861   Theological Seminary
8 September 1861   Fairfax Theological Seminary
Installments[180]   Theological Seminary, Virginia
- 23 September 1861
- 27 September
- 28 September
- 29 September
- 1 October
- 2 October
- 3 October
- 4 October
- 5 October
- 6 October
- 7 October
- 8 October
- 9 October
- 10 October
- 11 October

12 November 1861   Seminary Camp, Fairfax, Virginia
22 November 1861   Camp Seminary Virginia
4 December 1861   Camp Seminary
24 December 1861   Camp Seminary Virginia
6 February 1862   Camp Seminary Virginia
27 February 1862   Camp Seminary Virginia
29 April 1862   York County Virginia
30 April 1862   Cedar Creek & Catlett Station
8 May 1862   On land near York River
Installments
- 11 May 1862   White House Pamonkey River[181]
- 17 May   White House Pamonkey River
- 18 May

- 19 May	Near Kent County
- 20 May
- 22 May
- 23 May
- 24 May	Hanover County, Virginia
- 25 May	Hanover County, Virginia
- 26 May
- 27 May
- 28 May	29 May	Near Chicahominee River[182]
- 30 May
- 31 May	In camp

24 May 1862	Hanover County Virginia

In Gratitude

Writing this book took more than twenty years. It was an idea that had to percolate and simmer. For a long time, I did not realize what I had.

First, I applaud my mother Betty Robbins, grandmother Eleanor Sutton, and great-grandmother Vashti Shinn Sutton, for carefully preserving the letters in this collection and keeping their existence alive in family conversation.

Thank you to my brother Bill Robbins and his wife Karen Robbins, both members of the Mount Holly Monthly Meeting. They listened to me agonize over what would set any Civil War book I might write apart from the countless other scholarly volumes already out there. It was Bill who pointed out the obvious, that had been right under my nose all along: the inherent moral conflict for Friends who decide to go to war. As a Trustee of the Burlington Quaker Meeting House and Center for Conference, Bill helped me with access to burying ground records. Both Karen and Bill provide me with objective, well thought out observations.

Paulie Wenger of the Pemberton Township Historic Trust was helpful, searching maps to find the location of the Allinson farm. Historian Tica Simpson suggested paying more attention to the ladies. I had pretty much glossed over their sacrifice in this awful war. This is their story, too. Tad LaFountain reviewed the manuscript to make sure I portrayed the Religious Society of Friends accurately, and in a respectful manner. Many thanks to my dear, life-long friend Margaret Marks and her husband Ed Marks for their continued interest and encouragement.

Paul W. Schopp, Assistant Director of the Stockton University South Jersey Culture and History Center, agreed to meet me for coffee and look at what I had done. The outcome of that visit is what is before you now. He offered to be my editor and proposed that the book be published by the University. I was surprised, flattered, and delighted.

Thank you to Thomas Kinsella for shepherding the book to print.

Works Cited

Books
Banquet, Camille, *History of Kearny's First New Jersey Brigade* (1910), published by the State of New Jersey, Trenton.

Bilby, Joseph G., and Goble, William C., *"Remember You Are Jerseymen!" A Military History of New Jersey's Troops in the Civil War*, (Hightstown, NJ: Longstreet House, 1998).

Buck, James G., *Journey to Honor* (College Station, TX: Virtualbookworm.com Publishing, 2008).

Cullen B. Aubrey, *Recollections of a Newsboy in the Army of the Potomac, 1861–1865. His Capture and Confinement in Libby Prison, after Being Paroled, Sharing the Fortunes of the Famous Iron Brigade* (Milwaukee, Wisconsin, 1904). Available at babel.hathitrust.org/cgi/pt?id=n-jp.32101046794127&seq=1.

Garrison, Nancy Scripture, *With Courage and Delicacy, Civil War on the Peninsula, Women and the U.S. Sanitary Commission* (Mason City, Iowa: Savas Publishing Company, 1999).

Gottfried, Bradley M., *Kearny's Own: The History of the First New Jersey Brigade in the Civil War* (New Brunswick, NJ: Rutgers University Press, 2005).

Jackson, William J., *New Jerseyans in the Civil War* (New Brunswick, NJ: Rivergate Books of Rutgers University, 2006).

Jones, Rufas, *The Quakers in the American Colonies* (New York: W. W. Norton & Company, Inc., 1966).

Miller, William J., *The Men of Fort Ward* (Alexandria, VA: Published by the Friends of Fort Ward, 1989).

Philadelphia Yearly Meeting, *Faith and Practice* (Philadelphia, PA, 2011).

Phisterer, Frederick, *Statistical Record of the Armies of the United States* (New York: Charles Scribner's Sons, 1883).

Swinton, William, *Campaigns of the Army of the Potomac* (New York: Charles Scribner's Sons, 1882; revision and re-issue of 1866).

Thompson, Robert L., *Burlington Biographies: A History of Burlington, New Jersey Told Through the Lives and Times of Its People* (Galloway, NJ: South Jersey Culture & History Center, 2016).

Journals
The American Annual Monitor for 1859; or Obituary of the Members of the Society of Friends in America for the Year 1858, volume 2.

Jenson, Anika, *The Oatmeal Brigade: Quaker Life During the Civil War* (The Gettysburg Compiler, 2015).

The Sanitary Commission of the United States Army: A Succinct Narrative of its Works and Purposes (New York: USSC, 1864).

Newspapers
Aubrey, Cullen, "Reflections of a Newsboy in the Army of the Potomac," *The Philadelphia Inquirer*, September 25, 2009.

Courier-Post (Camden, NJ), May 4, 1934.

Daily Record (Long Branch, NJ), September 13, 1929.

Hartford Courant (Hartford, CT), July 31, 1811.

Lancaster Intelligencer (Lancaster, PA), Feb. 19, 1908.

Messenger-Press (Allentown, NJ), May 10, 1928.

Monmouth County Democrat (Freehold, NJ), May 22, 1862.

Morris County Chronicle (Morristown, NJ), May 28, 1908.

Mount Holly News (Mount Holly, NJ), February 18, 1908.

New Jersey Mirror (Mount Holly, NJ), February 15, 1855, September 28, 1858, February 19, 1908.

The Philadelphia Inquirer (Philadelphia, PA), February 15, 1908.

Trenton Evening Times (Trenton, NJ), May 1, 1899.

West-Jersey Pioneer (Bridgeton, NJ), May 17, 1862.

Websites

Gigantino, Anthony (2019) "The Philadelphia Quakers in the Civil War," *The Histories* Vol. 4: Iss. 2, Article 4, Available at: https://digitalcommons. lasalle.edu/the_histories/vol4/iss2/.

"Balloons in the American Civil War," *U.S. Centennial of Flight Commission*, www.centennialofflight,gov/essay/Lighter_than_air/Civil_War_balloons.

"Bailey's Cross Roads Advanced Post of the United States Army Opposite Munson's Hill" www.sonofthesouth.net/Ieefoundation/civil-war/1861/october/baileys-cross-roads.htm.

"A Brief History of the Branches of Friends," Quaker Information Center, https//quakerinfo.org/Quakerism/branches/history.

"Bristoe Campaign, American Battlefield Trust, " http://bristoe.lstminnd. org/.

"Civil War Military Bands," American Battlefield Trust, www.battlefields. org/learn/articles/battle-bands.

"Civil War Soldiers and Sailors," National Park Service, http://www.itd. nps.gov/cwss/soldiers.cfm.

"Civil War," United States Postal Service, usps.com.

"Clearness Committees" Friends General Conference, FGCquaker.org.

"The Cooper Family Historical Marker," The Historical Marker Database, The Cooper Family Historical Marker, hmdb.org

Farrow, Edward Samuel, "Farrow's Military Encyclopedia, a Dictionary of Military Knowledge, Vol. 2," Internet Archive.

"The First Air Forces – A Century of Balloons at War," Military History Now, https://militaryhistorynow.com/2012/07/05/early-air-power-100-years-of-balloons-at-war/.

"Fort Ward Museum and Historic Site," http://oha.alexandtiava.gov/fort-ward/fw-history.html

The Grundy Archive," Margaret R. Grundy Memorial Library, Bristol, PA.

"History of the Meetinghouse" Burlington, NJ Quaker Meeting House & Center for Conference | History (burlmhcc.org)

Lewis, Justin, "Quakers in the Civil War," Cathedral of Liberty, Quakers in the Civil War – Cathedral of Liberty (cthl.org).

Library of Congress, www.loc.gov.

"Medicine in the Civil War," Wikipedia.

"Money, Money, Money!" —19th Century Currency— Soldier Pay in the American Civil War Activity," themua.org.

"Monthly Wage of Soldiers in the American Civil War from 1861 to 1865, by Rank," Statistica, https://www.statista.com/statistics/1032399/wage-rank-american-civil-war-1861-1865.

"New Jersey Biographical Sketches 1685–1800," Ancestry.com.

"New Jersey in the Civil War" https://americancivilwarinstitute.blogspot.com/2013/08/new-jersey-in-civil-war.html.

"The Creation of the U.S. Sanitary Commission," Lawrence Weber, February 2010, https://warfarehistorynetwork.com.

"The Mud March" U.S. National Park Service, nps.gov.

"Roebling . . . Wire Rope and American Bridges," The Historical Marker Database, www.hmdb.org.

Seliga, Joseph Francis, "The Search for Camp Olden Hamilton Township," http://www.campoldemorg/researchBody.html, May 20, 1995.

"Soldiers' Pay," New Market Historical Society, /www.newmarketnhhistoricalsociety.org.

United States Postal Service, "Sources of Historical Information on Post Offices, Postal Employees, Mail Routes, and Mail Contractors," Publication 119.

"U.S., Civil War Pension Index: General Index to Pension Files, 1861-1934," Ancestry.com.

Archives and Special Collections
Burlington County Clerk's Office, Mount Holly, New Jersey, Deed book D.

Burlington Religious Society of Friends Monthly Meeting minutes
New Jersey Civil War records are available on www.njstatelib.org/NJ-ln-formation.

Monmouth County Clerk's Office, Freehold, New Jersey (monmouth-countyclerk.com).

New Jersey, Wills and Probate Records, 1656–1999.

The New York Public Library Manuscripts and Archives Division Guide to the United States Sanitary Commission records, https://archives.nypl. org/.

Rutgers University. Libraries. Special Collections and University Archives, Rutgers University. Libraries. Special Collections
> David Allinson's papers are stored at Rutgers University in the Special Collections Repository, Website: ALLINSON, David, Burlington | Archives and Special Collections at Rutgers.

Interviews and letters
Allison, Joseph, original letters 1853–1862.

Letter to Mrs. Aaron S. Robbins, May 3, 1863, from Master Sergeant, 11 Rgt. Co. E., in Robbins family archives.

Morris, Rebecca, historian, expert in Civil War parole camps, Ann Arundel County (MD) Historical Society.

148

Bibliography

Ancestry.com

Banquet, Camille, *History of Kearny's First New Jersey Brigade* (Trenton, NJ: published by the State of New Jersey, 1910.

Bilby, Joseph G, and Goble, William C., *"Remember You Are Jerseymen!" A Military History of New Jersey's Troops in the Civil War* (Hightstown, NJ: Longstreet House, 1998).

Buck, James G., *Journey to Honor* (College Station, TX: Virtualbookworm.com Publishing, 2008).

Cox, Christopher, *History of New Jersey Civil War Regiments: Artillery, Cavalry, and Infantry* (Middletown, DE, 2013).

Dictionary of Quaker Biography, Biographical Sketches in Typescript (Special Collections, Haverford College Library).

Fritz, Peter; Naylor, Ron; Kaewell, Edwin, *A History of the N.J. Vol. Infantry* (Published by the Delanco Historic Preservation Advisory Board).

Gottfried, Bradlely M., *Kearny's Own: The History of the First New Jersey Brigade in the Civil War* (New Brunswick, NJ: Rutgers University Press, 2005).

History of the Reunion Society of the 23rd Regiment, N.J. Volunteers (Philadelphia: Keystone Printing Co., 1890).

Haskew, M. E., "Prisons of the Civil War: An Enduring Controversy," Warfare History Network, 2013.

Jackson, William J., *New Jerseyans in the Civil War* (New Brunswick, NJ: Rivergate Books of Rutgers University, 2006).

Martz, Susan, *Discover Beverly Edgewater Park New Jersey* (Cincinnati, OH: The Creative Company, 1993).

Massey, Mary Elizabeth, *Women in the Civil War* (New York: A. A. Knopf, 1966).

Miller, William J., *The Men of Fort Ward, Friends of Fort Ward* (Alexandria, VA, 1989).

Newspapers.com.

Pomona Hall: A Quaker Plantation of the Eighteenth Century (Camden, NJ: Camden County Historical Society, 1986).

Pickenpaugh, R. (n.d.). "Prisoner Exchange and Parole." ESSENTIAL CIVIL WAR CURRICULUM, www.essentialcivilwarcurriculum.com.

Rodens, J. P., *The Fighting Quakers* (New York, 1886; reissued by A. J. H. Duggane, Digital Ninja Media, Inc.)

Scott, Robert N., *The War of the Rebellion: A Compilation of Official Records of the Union and Confederate Armies* (Washington, DC.: prepared under the direction of the Secretary of War, Government Printing Office, 1885).

Talavera, Dorothy Robbins, *The Civil War Letters Written to Members of the Shinn and Allinson Families of Burlington County, New Jersey* (Delanco, New Jersey, 2008).

Talavera, Dorothy Robbins, "Sons of Burlington County Rush to Save the Union," *Beverly Bee* (Beverly, NJ), June 2010, 28.

Talavera, Dorothy Robbins, "From the Naval Archives: Letters from Private William E. Shinn, 1864–1866," *Civil War Navy – The Magazine* Vol. 13, Issue 3 (Winter 2026), 61.

Thompson, Robert L., *Burlington Biographies: A History of Burlington, New Jersey Told Through the Lives and Times of Its People* (Galloway, NJ: South Jersey Culture & History Center, 2016).

Woodward, E. M., *History of Burlington County, New Jersey* (1883; reprinted by Burlington County Historical Society).

About the Author

Dorothy Robbins Talavera is a proud Jersey Girl, who has reinvented herself several times over. First retiring from a career with the federal government in Washington, DC, she then returned to New Jersey and became a Spanish teacher. Now retired again, she immersed herself in the task of documenting her family history using archives in the family collection. She annotated and published her father's memoir, *The Life and Times of a One-Armed Surgeon, by Morris A. Robbins, MD*. Then, she wrote *Miss Sutton Does Things Differently*, the story of an aunt's remarkable achievements. In addition, she has published numerous articles in magazines, technical journals, and newspapers. There are several more such projects in the works.

Dorothy's late husband Jose Miguel Talavera-Toso was a Peruvian immigrant who joined the United States Army as a teenager to fight in Vietnam. He returned a highly decorated and badly damaged man, who helped other veterans for the rest of his life. He and Dorothy raised three beautiful daughters.

152

Endnotes

1   The terms Quaker, Religious Society of Friends, and Friends are used interchangeably throughout this story. They are all one and the same. See Glossary for explanation.

2   Cooper Historical Markers located at 206 and 326 Cooper Streets, Camden, New Jersey, near Rutgers University.

3   Robert L. Thompson, *Burlington Biographies: A History of Burlington, New Jersey, Told Through the Lives and Times of Its People* (Galloway, New Jersey: Stockton University, 2016), 163.

4   Rufas Jones, *The Quakers in the American Colonies* (New York: W.W. Norton & Company, Inc., 1966), 457.

5   Thompson, op. cit., 161.

6   Friends House library and Haverford College, *Digital Dictionary of Quaker Biography* (archive.org: https://ia600209.us.archive.org/view_archive. php?archive=/18/items/wiki-trilogybrynmawredu_speccoll_dictionary/ trilogybrynmawredu_speccoll_dictionary-20140324-wikidump.7z&- file=index.html

7   Allinson family papers (Brynmawr, PA: Brynmawr College, Archives & Manuscripts): brynmawr.edu.

8   David Allinson papers, Elmer T. Hutchinson Collection of Biographical and Bibliographical Notes on Pre-1850 New Jersey Printers and Publishers, Elmer Tindal Hutchinson Papers (MC 866) (New Brunswick, NJ: Rutgers University, Alexander Library, Special Collections) https:// archives.libraries.rutgers.edu/.

9   Burlington County Deed Book D (Mount Holly, NJ: Burlington County Clerk's Office, 1795), 216.

10 *The Philadelphia Inquirer* (Philadelphia, PA), July 5, 1823, 4.

11 According to the Cambridge University Press, an economic crisis occurred in 1816–1817 that affected the entire Western world, the result of European upheaval stemming from the Napoleonic wars and a climate catastrophe that caused fluctuation in agriculture commodities. This may have contributed to Allinson's misfortune. *The Journal of Economic History*, 248, published online by Cambridge University Press, https://www.cambridge.org/core/journals/journal-of-economic-history/ article/abs/economic-crisis-of-18161817-and-its-social-and-political- consequences.

12 *The American Annual Monitor for 1859, or Obituary of the Members of the Society of Friends in America, for the year 1858, No. 2* (New York City, NY: Samuel S. & William Wood, 1859).

13 Beaulah Zane appears on the 1815 membership roll of the Old Saint George Methodist Episcopal Church, Philadelphia. Her parents' names appear in the Abstract Record of Births, Deaths, and Burials of the Philadelphia Monthly Meeting, 1688–1826.

14 State and Federal Census records accessed through Ancestry.com.

15 "U.S., U.K., and Ireland Quaker Published Memorials for 1818–1819…," *The American Annual Monitor for 1859*, Vol. 2 (New York City, NY: Tract Association of Friends, 1859), 10.

16 In 2024, an online auction offered a copy of *The Scrivener's Guide* by William Griffith. David Allinson, Burlington, New Jersey, printed the work in 1813. Bidding began at $170. *Rare antique 1813 Scrivener's Guide early law book #510037725,*www.worthpoint.com/worthopedia.

17 *New Jersey Mirror* (Mount Holly, NJ), September 28, 1858, 1.

18 Will Book I-K 1857–1866 (Mount Holly, NJ: Burlington County Surrogate's Office), 195.

19 In most documents and some (but not all) newspaper articles, Census data, and New Jersey Civil War military records, the surname appears as "Allinson," but Joseph signs his own name and writes to members of his family as "Allison."

20 *New Jersey Mirror* (Mount Holly, NJ), February 8, 1855, 3.

21 *New Jersey Mirror* (Mount Holly, NJ), February 15, 1855, 3.

22 Joseph's mother Beulah was the sister of Vashti's grandmother Abigail Zane.

23 John Cooper Allinson, Joseph's brother.

24 May be his cousin, Samuel Allinson.

25 David Cooper Allinson, Joseph's brother.

26 "Death Claims D.C. Allinson at Burlington," *The Times* (Trenton, NJ), February 14, 1908, 1.

27 Ibid.

28 Ibid.

29 *Daily Record* (Long Branch, NJ), Sept 13, 2029.

30 "Edgebrook," *Messenger Press* (Allentown, NJ), May 10, 1928, 1.

31 *Mount Holly News* (Mount Holly, NJ), February 18, 1908, 1.

32 Letter from Joseph Allison to his mother from Camp Seminary Virginia, November 22, 1861

33 *Trenton Evening Times* (Trenton, NJ), May 1, 1899, 1.

34 *Sources of Historical Information on Post Offices, Postal Employees Mail Routes, and Mail Contractors*, United States Postal Service Publication 119 (Washington, DC: Government Printing Office, 2011).

35 Caroline Foster, born 1850. Her father, Asa Robbins Foster, died when

she was 11; her mother died soon after Carrie's marriage. She was related to Beulah Allinson. Joseph Allison's letters often mention Uncle Asa and Aunt Harriet. Carrie married Thomas Hattersley in 1870. They named their firstborn child Allinson Isaac Hattersley.

36 Children of John and Lucy.

37 Will Book W-X, 1907–1910 (Trenton, NJ: Mercer County Surrogate's Office, February 26, 1907), 271.

38 *The Philadelphia Inquirer* (Philadelphia, PA), February 15, 1908, 4.

39 Ibid.

40 *Mount Holly News* (Mount Holly, NJ), February 18, 1908, 1.

41 Nephew Charles E. Allinson was the individual Cooper defended in court when his rent payments entered arrears.

42 *New Jersey Mirror* (Mount Holly, NJ), February 19, 1908, 3.

43 J. D. Scott, *Combination Atlas Map of Burlington County, New Jersey*, Pemberton Township plate (Philadelphia, PA: J.D. Scott, 1876), 40–41.

44 Burlington Monthly Meeting (Philadelphia Yearly Meeting), Marriages records, 1768–1897, 264–65.

45 Upper Springfield Monthly Meeting, Record of Members, 1783–?, Births and Deaths 1707–1833, No. 541, 83.

46 This would equate to more than a $250,000 today.

47 William J. Jackson, *New Jerseyans in the Civil War* (New Brunswick, NJ: Rivergate Books, Rutgers University Press, 2006), 35.

48 Ibid, 37.

49 *Faith and Practice* (Philadelphia, PA: Philadelphia Yearly Meeting, 2001), 76.

50 Ibid, 75.

51 Ibid.

52 Minutes of the Monthly Meeting held in Burlington from the 12th month 1802 to the 2nd month 1820, 334.

53 Anthony Gigantino, "The Philadelphia Quakers in the Civil War," *The Histories,* Vol. 4, No. 2, Article 4, 2019. Available at: https://docslib.org/doc/1777491/the-philadelphia-quakers-in-the-civil-war-by-anthony-gigantino.

54 Anika Jenson, "The Oatmeal Brigade: Quaker Life During the Civil War," (2015), www.gettysburgcompiler.org.

56   Faith and Practice, Philadelphia Yearly Meeting, 78. https://www.pym.org/faith-and-practice/.

57 New Jersey Civil War Records, New Jersey State Library, www.njstatelib.org/NJ-information.

58 Cullen B. Aubrey, *Recollections of a Newsboy in the Army of the Potomac,*

1861–1865. His Capture and Confinement in Libby Prison, after b\Being Paroled, Sharing the Fortunes of the Famous Iron Brigade (Milwaukee, WI[?], 1904[?]), 138. Available at: https://babel.hathitrust.org/cgi/pt?id=njp.32101046794127&seq=1.

59 Letter from Joseph Allison to his mother Beulah Allison from Camp Seminary, November 22, 1861.

60 A list of casualties at Fort Ward appears in *The Men of Fort Ward* by William J. Miller (Alexandria, VA: Friends of Fort Ward, 1989), 45–47.

61 "Peculiar institution of slavery" is a euphemistic term in popular use in the South started in 1830 by southern politician John C. Calhoun.

62 Joseph Allison diary entry May 30, 1862, from Chickahominy.

63 American Battlefield Trust, "Gaines Mill," www.battlefields.org/learn/civil-war/battles/gaines-mill.

64 "Pickets are an outpost or guard for a large force. Ordered to form a scattered line far in advance of the main army's encampment, but within supporting distance, a picket guard comprised a lieutenant, 2 sergeants, 4 corporals, and 40 privates from each regiment. Picket duty constituted the most hazardous work of infantrymen in the field. Being the first to feel any major enemy movement, they were also liable to be killed first, or wounded, or captured. Snipers often targeted pickets. Picket duty, by regulation, was rotated regularly in a regiment." www.thomaslegion.net/americancivilwar/differencebetweenskirmisherandpicket.html

65 Fatigue duty is labor assigned to military men that does not require the use of arms according to Edward Samuel Farrow in *Farrow's Military Encyclopedia; a Dictionary of Military Knowledge*, vol. 2 (New York: Broadway, 1885), 623.

66 "The Killed and Wounded," *New Jersey Mirror* (Mount Holly, NJ), July 24, 1862, 3.

67 This author had several relatives in the Twenty-third Regiment.

68 "Beverly Two-Day Celebration Honors Its 150th Anniversary," *Courier-Post* (Cherry Hill, NJ), September 9, 2007), 16.

69 Letter from Aaron S. Robbins to his wife, September 14, 1862. Robbins was the great-great grandfather of this author. An article about Robbins called "Sons of Burlington County Rush to Save the Union" by Dorothy Robbins Talavera, appeared in the *Beverly Bee* (Beverly, NJ), June 2010, 26.

70 *West-Jersey Pioneer* (Bridgeton, NJ), August 9, 1862, 2.

71 Bradley Smith, *Soldiers' Pay*, Essential Civil War Curriculum, Virginia Center for Civil War Studies at Virginia Tech, 2010–2015. Available at: www.essentialcivilwarcurriculum.com.

72 "Soldier's' Pay," New Market (Virginia) Historical Society. Available at: https:nhnmhs.com/military/civil-war-2.

73 Muster Role, June 27, 1863, transcribed by ZJR, 5 March 2024 (Trenton, NJ: New Jersey State Archives), www.archives.nj.gov.

74 James G. Buck, *Journey to Honor* (College Station, TX: Virtual Bookworm Publishing, 2008), 11.

75 Henry Ogden Ryerson, son of a prominent New Jersey family, had worked his way up through the ranks from Private.

76 Joseph G. Bilby and William C. Goble, *Remember You Are Jerseymen: A Military History of New Jersey's Troops in the Civil War* (Hightstown, New Jersey: Longstreet House, 1998), 311.

77 Ibid.

78 Ibid, 314.

79 The inscription on the White Oak Church Historical Marker appears in the Appendix.

80 A list of deaths, as reported in the *New Jersey Mirror*, appears in the Appendix.

81 "The Mud March," National Park Service, Fredericksburg & Spotsylvania National Military Park. Available at https://nps.gov/articles/000/mudmarch/htm.

82 William Swinton, *Campaigns of the Army of the Potomac* (New York: Charles Scribeners and Sons, 1882), 260.

83 Bilby and Goble, op. cit., 315.

84 Camile Banquet, *History of Kearney's First New Jersey Brigade* (Trenton, NJ: State of New Jersey, 1910), 244.

85 Letter to Mrs. Aaron S. Robbins from Master Sergeant, 11 Rgt. Co. E., May 3, 1863. Letter in Robbins family archives.

86 "Local Facts and Fancies/The Burlington County Regiment," *New Jersey Mirror* (Mount Holly, NJ), May 14, 1863, 3.

87 "Correct List of the Killed, Wounded, and Missing," *New Jersey Mirror* (Mount Holly, NJ), May 21, 1863, 3.

88 Personal Communication: Rebecca Morris, historian and expert on Civil War parole camps, Ann Arundel County (Maryland) Historical Society, via email (April 3, 2024).

89 "Muster-Out Roll of Captain John P. Burnett's Company I in the Twenty-Third Regiment of New Jersey Infantry Volunteers commanded by Col. E. Burd Grubb called into service of the United States by the President at Beverly, New Jersey (the place of general rendezvous) on the first day of September 1862 to serve for the term of nine months from the date of enrollment unless sooner discharged; from the thirteenth day of

September, 1862, (when muster in) to the twenty-seventh day of June when mustered out. The Company was organized by John P. Burnett at Mount Holly N.J., in the month of August 1862 and marched there to Beverly, when it arrived the (blank) day of September, a distance of (blank) miles." (Trenton, NJ: New Jersey State Archives, Adjutant General files, 1863).

90 Buck, op. cit., 217–18.

91 United States Civil War Pension Index: General Index to Pension Files, 1861–1934, for John C Allinson, found on Ancestry.com.

92 Peter Fritz, Ron Naylor, and Edwin Kaewell, *A History of the N.J. Vol. Infantry* (Delanco, New Jersey: Delanco Historic Preservation Advisory Board).

93 Veteran Aaron S. Robbins made this presentation in Hedding's one-room schoolhouse in Mansfield Township, Burlington County. Little Sarah Poinsett, a student, would grow up to marry Aaron's grandson, Clarence Robbins, this author's grandparents. Sarah recounted this story to her grandchildren many times.

94 *Courier-Post* (Camden, NJ), May 4, 1934, 13.

95 *The Mount Holly News* (Mount Holly, NJ), February 18, 1908, 1.

96 John Davis Billings, *Hardtack and Coffee, or the Unwritten Story of Army Life* (Boston, MA: George M. Smith & Co., 1887), 298. Available at: https://archive.org/details/hardtackcoffeeor00bill.

97 Lawrence Weber, *The Creation of the U.S. Sanitary Commission*, Warfare History Network (2010). Available at: https//warfarehistorynetwork.com/article/the-creation-of-the-u-s-sanitary-commission/.

98 Nancy Scripture Garrison, *With Courage and Delicacy, Civil War on the Peninsula, Women and the U.S. Sanitary Commission* (Mason City, IA: Savas Publishing Company, 1999), 1.

99 Ibid, 6.

100 Guide to the United States Sanitary Commission Records (The New York Public Library Manuscripts and Archives Division), https://archives.nypl.org/.

101 *The Sanitary Commission of the United States Army: A Succinct Narrative of its Works and Purposes.* (New York: United States Sanitary Commission, 1864), 231.

102 "Death Claims D. C. Allinson at Burlington," *The Times* (Trenton, NJ), February 14, 1908, 1.

103 *Lancaster Intelligencer* (Lancaster, PA), February 19, 1908, 1.

104 Letter to "Sister" from Joseph Allison, written at Camp Seminary, VA, December. 4, 1861.

105 Letter to "Sister" from Joseph Allison, written at Camp Seminary, VA, February 6, 1862.

106 Letter written to "Mother" from Joseph Allison, written at Camp Seminary VA, November 22, 1861.

107 "At Pemberton, on the 1st instant (March 1863), Lucy, daughter of John C. and Lucy A. Allinson, aged 10 weeks and 3 days." *New Jersey Mirror* (Mount Holly, NJ), March 19, 1863, 3.

108 *Monmouth Democrat* (Freehold, NJ), Feb 16, 1871, 3.

109 *The Mount Holly News* (Mount Holly, NJ), March 23, 1897, 3.

110 *The Times* (Philadelphia, PA), May 3, 1897, 5.

111 William F. Newbold, Map of Friends' Burying Ground, Burlington, New Jersey, manuscript, 1820, updated March 1975 by H. S. Haines (Burlington Quaker Meetinghouse and Center for Conference).

112 Frederick Phisterer, *Statistical Record of the Armies of the United States*, (New York City, NY: Charles Scribner and Sons, 1883), 14.

113 Buck, op. cit., 8.

114 Joseph Francis Seliga, *The Search for Camp Olden Hamilton Township*, (Hightstown, NJ, Longstreet House, 1995). Available at http://www.campolden.org.

115 National Park Service, Soldiers and Sailors Database, https//www.nps.gov/civilwar/search-battle-units.htm.

116 Known today as Virginia Theological Seminary; during the Civil War, Union troops occupied its grounds, and its buildings served as a military hospital.

117 Joseph was writing to his cousins, the family of Earl and Emma Arey Shinn of Mansfield, New Jersey.

118 The troops initially were issued obsolete smoothbore Springfield muskets used to fill the requirement for arms.

119 Barclay White (1821–1906), was a well-known Quaker and pioneer New Jersey cranberry farmer, historian, and author. Newspapers published his work on local history. After the war, the president appointed him Superintendent of Indian Affairs.

120 Asa Robbins Foster and his wife Harriet of Springfield Township, New Jersey, were parents of Caroline P. Foster Hattersley, whom David Cooper Allinson adopted after their death.

121 William J. Miller, *The Men of Fort Ward* (Alexandria, VA: Friends of Fort Ward, 1989), 3.

122 Fort Ward Museum and Historical Site (Alexandria, VA: Historic Alexandria | City of Alexandria, VA). Map found at NPS.GOV.

123 Quotation from English essayist, playwright and politician Joseph

Addison (1672–1719) in his *Spectator* magazine.

124 Adjutant General William S. Stryker's two-volume work, *Record of Officers and Men of New Jersey in the Civil War, 1861–1865*, held in the New Jersey State Archives collections, suggests this might have been David H. Brower, a 28-year-old musician from Rahway, New Jersey, who died of typhoid on Sept. 6, 1861.

125 Vashti Burtis Shinn (later, Sutton) was the great-grandmother of this author. It is through her that these letters were compiled, saved, and passed down.

126 Whiskey played an important role in the war: social lubricant, relief of boredom, providing solace for loneliness, and a way to forget fear and horror. It was also used medicinally. Drunkenness was a problem and sometimes presented disciplinary problems. A delicate balance existed between morally appropriate consumption and excess.

127 Bradley M. Gottfried, *Kearny's Own: The History of the First New Jersey Brigade in the Civil War* (New Brunswick, NJ: Rutgers University Press, 2005), 30.

128 www.geocities.com/Yosimite/trails/9401/civilwar.html (webpage no longer available).

129 Now known as the 14th Street Bridge from Washington, D.C. to Arlington, VA.

130 The U.S. Army replaced the already obsolete rifles issued at Camp Olden, probably with the Springfield Model 1861, the most widely used U.S. Army weapon during the Civil War. The soldiers favored this rifle for its range, accuracy, and reliability.

131 Buck, op. cit., 69.

132 The Marshall House Hotel was at the corner of King and South Pitt street in Alexandria. Today, it is a Holiday Inn. A plaque on the exterior wall commemorates the incident.

133 Jarred Marlowe, *Ellsworth, Embalming, and the Birth of the Modern American Funeral,* Emerging Civil War, 2025. Available at https// emergingcivilwar.com/2025/07/24/ellsworth-embalming-and-the-birth-of-the-modern-american-funeral/

134 Gottfried, op. cit., 34.

135 Munson's Hill and Baileys Crossroads are approximately 2 miles from Alexandria. Today it is the intersection of Route 7 and Route 50.

136 There is no attempt to edit the language used in these letters. This term, although offensive today, was in common use at that time.

137 This quote appears to be from Ralph Waldo Emerson's 1841 poem, "Self Reliance." Like many people at that time, Allison memorized—and

often quoted—long passages of poetry.

138 More than 200,000 native-born Germans immigrants fought in the Civil War. They were the largest ethnic group among the troops. Albert Berhardt Faust, *The German Element in the United States . . .* (Boston, MA: Houghton Mifflin, 1909), 523, quoting from Benjamin Apthorp Gould, *Investigations in the Military and Anthropological Statistics of American Soldier* (New York City, NY: Published for the United States Sanitary Commission by Hurd & Houghton, Cambridge, Riverside Press, 1869). Available at: https://archive.org/details/bub_gb_Q9lCAAAA-IAAJ/mode/1up?q=Gould.

139 Mary Laine, *Woolson, Albert Henry (1850–1956),* Minnesota Historical Society Mnopedia, www3.mnhs.org/mnopedia (2019).

140 Two of Vashti's brothers did join the service. Joseph Shinn enlisted in the 3rd Reg In May 1861; William Shinn joined the Marines the same year.

141 American politician Patrick Henry's famous quote made in Virginia on Mar. 23, 1775. Allison's abolitionist grandfather, Samuel Allinson, may have influenced this quote after writing to Henry.

142 Joseph Arey Shinn, Vashti's brother, who already served in the 3rd Regiment.

143 Some in the New Jersey state legislature sought to build support for a last-ditch resolution to reach a peaceful solution to the crisis, involving a compromise on slavery to appease the southern states. They faced criticism as "traitors." Popular consensus strongly supported preserving the Union.

144 Balloons in the American Civil War (U.S. Centennial of Flight Commission, www.centennialofflight.net, 2013).

145 Roebling . . . Wire Rope and American Bridges, Historical marker located in Trenton, NJ, The Historical Marker Database, www.hmdb.org.

146 Allison's mother Beulah did, indeed, receive his pay after his death.

147 Allison used the Quaker form of address "thee" with his mother and sister-in-law, but not with members of the Shinn family. The Shinns were not Quaker.

148 He refers to his estranged brother, Cooper.

149 Allison's father David died in 1858. It appears there was a dispute between Joseph and his brother David Cooper (whom they called Cooper) about the execution of the will and estate settlement. "That scamp in Trenton" refers to Cooper.

150 Camp Olden is thought to have stood at the upper or northeastern end

of Camp Avenue in Hamilton Township near the state arsenal. More than 8,900 troops went through Camp Olden during the 193 days of its operation. A park at the corner of Liberty Street and Hamilton Avenue memorializes the camp. Matt Fair, "Civil War 150th Anniversary highlights Mercer County's historical ties," *The Times* (Trenton, NJ), April 17, 2022, www.nj.com/mercer.

151 Aaron O'Neill, Monthly wage of soldiers in the American Civil War from 1861 to 1865, by rank, Statistica, www.statista.com/statistics/1032399/wage-rank-american-civil-war-1861–1865, (2024).

152 William H. Johnson, Co. B, 1st NY Cavalry, shot for desertion Dec. 13, 1861, at Fairfax, VA. It was the first execution within the Army of the Potomac. Several accounts and illustrations can be found online and in books.

153 Jonathan Fox of Pemberton, NJ, died in December 1861 at age 100.

154 Sutler: An appointed army provisioner who supplied the soldiers with food and other necessities.

155 Provost guard is a military police detachment working for the Provost General. Allison was put in the guard house as punishment because he disobeyed orders and wandered outside the allowed camp perimeter. Apparently, jail proved more comfortable than his tent.

156 Allison may have authored some of the poetry that appears in his letters. I can find no source for the verses he pens here.

157 C. Siegel, "The Orange & Alexandria Railroad," E 2012, losthistory.net/csiegel.

158 John Greenleaf Whittier's poem "Ein Feste." The Allinson family knew Whittier.

159 Poquoson River.

160 Major Charles Sitgreaves (1803–1878) failed to be elected Governor. He did, however, serve as the American Democratic Party representative for New Jersey's 3rd congressional district to the state legislature from 1865 to 1869.

161 The author recognizes the offensiveness of this term but regretfully chooses to include it when transcribing the letters verbatim.

162 Postal censorship of soldiers' letters occurred on both sides during the war.

163 "Secesh" was the derogative term used for the Confederate soldiers

164 "Peculiar institution of slavery" is a euphemistic term in popular use in the South started in 1830 by southern politician John C. Calhoun.

165 *The New York Times*, May 24, 1862, 1, reported that the Rebel Cavalry had advanced beyond Turnstall's Station, within 14 miles of Richmond.

The article reported casualties and prisoners taken.

166 Possibly refers to Berdan Sharpshooters. Their uniforms were of fine material consisting of dark green coats and caps with black plumes, light blue trousers, and leather leggings, presenting a striking contrast to the regular blue uniforms of the infantry. The knapsack was of hair-covered calfskin, with a cooking kit attached, considered the best in use. Source: Berdan Sharpshooter History, http://www.berdansharpshooter.org.

167 *Monmouth Democrat* newspaper was published in Freehold, New Jersey from 1834 to 1942.

168 "Jerseymen in the Late Battle," *New Jersey Mirror* (Mount Holly, NJ), July 10, 1862, 3.

169 *New Jersey Mirror* (Mount Holly, NJ), September 18, 1862, 3.

170 *New Jersey Mirror* (Mount Holly, NJ), October 9, 1862, 3.

171 www.nps.gov/civilwar/soldiers-and-sailors-database.htm.

172 wwwnet-dos.state.nj.us/DOS_ArchivesDBPortal/StrykerCivilWar.aspx.

173 "Correct List of Killed, Missing, and Wounded," *New Jersey Mirror,* Mount Holly, NJ, May 21, 1863, 3.

174 Ibid.

175 A brevet is a warrant giving a commissioned officer a higher rank title as a reward for gallantry or meritorious conduct. It may not confer the authority, precedence, or pay of real rank, and is often granted just before retirement.

176 The Historical Marker Database, http.//www.hmdb.org.asp.

177 Definitions come from *Faith & Practice*, and copied from the website of the Philadelphia Yearly Meeting, Quakers & Quakerism – Philadelphia Yearly Meeting (pym.org).

178 Clearness Committee, Friends' General Conference, (Philadelphia, PA), FGCquaker.org.

179 Burlington, New Jersey, Quaker Meeting House & Center for Conference | Where Friends Have Gathered For Over 300 Years (burlmhcc.org).

180 The installments may have been intended as diary entries to be sent later for publication in the *New Jersey Mirror*. Allison asked sister-in-law Lucy Allinson to keep track and inform him of the number of the last "memorandum."

181 Allison's spelling; present day name is Pamunkey River.

182 Allison's spelling; present day name is Chickahominy River.

Index

Clevenger, Hannah, 92
Clevenger, Henry, 66, 72
Clouds Mill, 62, 66, 80, 81
Combs, C. H., 87
Cooper, David, 11
Cooper, Martha, 12, 13
Cooper, William, 11, 12, 15
Cox, Jason, 74
Cox, Joseph, 87
Cox, William, 74, 120
Croshaw, John, 87
Croshaw, Joseph, 87

Davis, Jeff, 62

Edsel, Mr., 72
Egbert, Frank, 87
Ellis, Peter, 93
Ellsworth, Col. Elmer, 66, 67
Emmons, John, 19
Emmons, George, 92, 120
Emmons, John, 19
Ewan, William M., 17

Fairfax Court House, 61, 95, 101
First New Jersey Brigade, 41, 44
First Regiment New Jersey Infantry, 54, 96
Flood, James, 38, 39, 120, 121
Forts
    Albany, 62
    Ellsworth, 66, 67, 69, 101
    Runyon, 62, 63, 64, 67
    Sumter, 33, 53
    Taylor, 56, 75, 77
    Ward, 9, 56, 57
Fort, A. H., 87
Fortress Monroe, 55, 62, 99, 108
Foster, Asa Robbins and wife Harriett, 24, 56, 62, 63, 67, 75, 92, 96, 99, 103
Fourth New Jersey Regiment, 71, 80, 88

www.ingramcontent.com/pod-product-compliance
Lightning Source LLC
Chambersburg PA
CBHW050000040726
47599CB00014B/1155